HUMAN ECOLOGY FOR GLOBALIZATION

HUMAN ECOLOGY
in ACTION

VOLUME – 1

Dr. (Mrs.) SHASHI KUMAR

Foreword by

Dr. PANJAB SINGH

SECRETARY
Deptt. of Agricultural Research & Education
and
DIRECTOR GENERAL
Indian Council of Agricultural Research (ICAR)

ATLANTIC
PUBLISHERS & DISTRIBUTORS

Published by
ATLANTIC PUBLISHERS AND DISTRIBUTORS
B-2, Vishal Enclave, Opp. Rajouri Garden, New Delhi-27
Phones : 25413460, 25429987, 25466842

Sales Office
7/22, Ansari Road, Darya Ganj, New Delhi-02
Phones : 23273880, 23285873, 23280451
Fax : 91-11-23285873
web : www.atlanticbooks.com
e-mail : info@atlanticbooks.com

ISBN 81-269-0226-4 (Vol. I)
ISBN 81-269-0227-2 (Vol. II)
ISBN 81-269-0147-0 (Set)

Typeset at
APD Computer Graphics, Delhi
Printed in India at
Nice Printing Press, Delhi

Secretary,
Department of Agricultural Research & Education
and
Director General,
Indian Council of Agricultural Research,
Krishi Bhawan,
New Delhi-110 001

Foreword

Human activities are changing the natural balance of the earth. Increasing the agricultural production for development in isolation is not the only issue of sustainability. To safeguard our biodiversity, we should earn grass-roots support. Scientists/ teachers who have established their excellence are required to take appropriate action at all levels for a change in human values and behaviour towards the holistic environmental ethics for better and safe future of mankind. It requires global co-operation to achieve the goal of sustainable development not only in agriculture but also in the entire national economy through various community development programmes for improving family living conditions.

Although the 90 indicators of human development neglect various dimensions of level of living, still human development such as literacy and schooling, health and longevity, employment, wage scrutiny creates a comprehensive data for interpretation, in rural and urban areas. Therefore, we must act fast to bring in right institutions in place for meeting the eco-challenges and access to knowledge for enhancing the capability and capacity in organization and management of biodiversity and food security through institutions of higher training.

I am sure, this publication would be a useful reference document for the students, professionals and policy makers.

(Panjab Singh)

Preface

The science of human ecology lies at the very heart of relationships between human being and the natural world. It is more than a scientific discipline. It is a philosophical study of Ecology, a way of looking at the world that sees connections among plants, animals and processes.

Worldwide force of globalization is causing deep change in many sectors of development. Free horizontal mobility can be facilitated through policies accompanied by adequate training facilities, both in number of persons trained and extent of training, for effective acquisition and mastery of foreign technology. Building up an effective research and development system, for linking imported technology with domestic research and development system, should earn grass-root support for effective acquisition. The study analyses close relationship with the force of virtualization in producing deep change among farm, off farm and non-farm sectors, with an enterprise by research institution.

I express my indebtedness to Indian Council for Agricultural Research, New Delhi and University Grants Commission, New Delhi for providing me sincere help and constant encouragement at all stages of my study.

It gives me immense pleasure to thank Horticulture and Education division NARS and NATP for encouragement and guidance during the course of development of the field of Human Ecology and expressing their views from time to time, which enabled me to complete these two Volumes. I am also grateful to Jawahar Lal Nehru University, especially, to premit me to include many recent developments to help development of our knowledge in this field which has been expanding at an explosive rate. I express my deep gratitude to Mr. H.C. Jain, Scientist—CSIR, and Mr. V.K. Saxena, Director—BSI, New Delhi for providing valuable help.

Last but not the least, I am thankful to M/s Atlantic Publishers and Distributors, New Delhi, particularly to Dr K.R. Gupta, its Managing Director, for bringing out the book in a record time.

SHASHI KUMAR

CONTENTS

VOLUME II

1

Introduction

While examining man-man relationship (needs and participation) and one aspect of man-nature relationship (the environment), the eco-development system has been emphasised on income generation for women. Environmental compatibility, one of the eco-development concerns has been reasonably achieved by carrying out income generating projects. Thus, by providing employment opportunities to women, a chain of development activities were initiated, the fruits of which can be observed over a period of time. Beside this, self-employment opportunities were developed for the women and marketing awareness was also created.

The basic mechanisms by which the technologies were transferred to the beneficiaries were (1) developing need based projects, (2) implementing the projects through trained personnel and (3) constant monitoring at each stage and immediate rectification of any problem arising during implementation. This mechanism was adopted as what is needed primarily is the development of systems for transference of technologies in their broadest sense. This includes the know-how technique, in a manner which promotes the recipient's own ability to become, her or himself, a technologist. Those who do not have access to technological knowledge or experience are never likely to become developers of their own technological solutions as those who do.

Any technological innovation brings with it a host of new problems which necessitate further research, both biological and social. The farmer needs the appropriate technology and the recommended inputs for increasing production. The farm trials laid out in the farmer's field and the field visits made jointly by the research and extension sub-systems bring the extension research and client sub-systems close together. The transfer of income generating technology related activities has profound implication in such fields as energy, environment, science and technology.

The success of the eco farming system approach lies in the development of technological packages which are realistic and relevant for the target group. It should recognize both the socio-economic background of peasant farmers and external circumstances imposing certain production and consumption condition of rural societies. New production methods which are consistent with peasant farmers' resources and internal external circumstances should be developed.

Besides this, efforts be made to study the level of technology to be transferred before actually transferring the technology. It was observed that less sophisticated technology which are developed locally are more appropriate and suitable to the target group than more sophisticated and imported technology which is transferred as such without taking into account the local situations prevalent. Very little or no 'research' has been conducted to determine the best farming systems that should be applied in a specific situation or system for *e.g.* pastoral farming systems are mostly replaced by sedentary farming systems in the west.

Farmers involvement is a fundamental step to a successful transfer of information and technology for sustainable land use. Sustainable land system in India refer to the successful management of resources to satisfy the increasing human needs in relation to agricultural livestock and energy, without degrading the environment or natural resources base or without compromising the needs of future generation. However, there is no methodology to develop and promote sustainable land use system. Besides this there is also no considerable experience available for the development and promotion of many of the technologies as social with sustainable system. The methodologies that evolved from farmer participatory research can definitely contribute to the development of a sustainable land use system research of farming system research.

The challenge is to consider the whole farm as a human activity system. This will require new mental, physical and operational tools and the need to understand the meaning of phenomena. Finally, we must pursue appropriate operationalising approaches in science and management. The new metaphor for creative action is in understanding the world to complement the current their chain of logic in a robust web of inter connections or network. The non quantifiable values must be reconciled in an operational landscape unit. In the new metaphor, the variable conditions of surroundings for a system will no longer be taken as controlled uniform or unchanging. The surroundings become a major feature of the management of a particular place.

Land use systems are dominated by small scale traditional farming that require the design of appropriate and improved technologies to meet the subsistence need of the family. However, water and soil fertility maintenance constitute the many existing constraints for all land use system.

Any multidisciplinary research and extension team that attempts to develop a methodology must first agree upon a common conceptual framework. The trade-off between short term economic viability and long term ecological sustainability include a systematic sequence of steps that over a period of time will result in the design, evaluation and promotion of new technology. These on integration, could evolve into more sustainable farming systems. The maintenance of natural resource productivity, more specifically, soil productivity is a pre-requisite and is tightly linked to plant production.

Introduction of more diverse cropping systems, especially involved combination of legumes, cereals, and perennial plants including trees can lead to substantial increases in soil resource use efficiency. These systems diversify the resourceful use of top soil nutrients and water from different soil depths. In order to improve the overall biological conditions of these soils, sub surface tillage was identified as the best kind which results in a higher level of organic matter. Integrated farming systems for farm activities have been developed. The farming systems cover mushroom farming, oil seed crop production, bee farming, pulse production and tuberose and gladiolus cultivation.

Basically the small Indian farmer is a reluctant and lazy farmer, village labour force moving to towns in the off-season and returning back in the sowing and harvesting season. This is a regular feature in the hill villages. Hence, even if the transfer of technology was done at the grass-roots level and man to man needs were tried to be achieved, it is not necessary that the farmers will continue with the process. This basic nature is thus a constraint in ensuring their further participation, once the projects ended and the inputs were withdrawn. Although, by this stage the farmers were trained to generate the inputs as well as channelise the marketing, they did little to further the work. Whilst the same was being managed by the project, it was fine with them.

Social and economic development must go hand in hand. An incorrigible society cannot secure a viable economy. In the context of the rural development it can be said that the philosophy of "extensiveness" must yield place to "intensiveness and consolidation". The country is poor economically and exploited for its development. Keeping this in view integrated system having many components were developed in the mushroom farming as well as triple benefit, fuel yield and various floriculture activities and the concept of participation has been experimented with. The element of participation is difficult to discuss, because we have not had the relevant level of experience with it. It seems reasonable to say that some measure of success has been achieved and new lessons have to be learnt from this experience.

Findings of research scientists should not only be aimed at publication in scientific journals the data should be formulated to help in resource management and policy analysis. This calls for multidisciplinary approach, especially inclusion of socio-economics in biological based research programme.

The projects have bridged major communication gaps, and have succeeded in transfer of Science and Technology at grass-roots level to the women within their resources and environment. The findings/observations of the projects provide broad guidelines for government and NGOs for planning and implementation of projects at State and National levels. The finding can be utilised effectively by other research organisations to further enhance the work areas. The students also benefit, as the case studies provide a base for work towards achievement of higher education (M.Sc/Ph.D). On the international level the projects can serve as a model for planning, similar work in other developing countries.

Future plan of action on sustainable agriculture :

1. Synthesis of sub-systems of sustainable agriculture.
2. Synthesis of a sustainable farming system based on location specific need through on farm research.
3. Development of research methods for evaluating the farming system with regards to its biological environmental and social efficiency.
4. Development of methodology for monitoring the improved system over a long period of time.

Rural household and farming systems have become a prime focus of development assistance in recent years. Both in applied research and in project work a farm family perspective is being adopted to achieve successful agricultural development.

The farm is no longer viewed solely as a production unit. The focus has shifted to the rural household in its relation to its socio-economic and cultural environment. The rural family is seen as the central decision-making unit for the utilisation of existing development potentials and the application of innovations. Simultaneously, the rural household accounts for social stability in the society and for securing the material and non-material essential needs of its individual members.

The peasant household is seen as the central decision-making unit in rural development. This refers not only to utilisation of existing development potentials, the implementation of development measures or the introduction of innovation of various kinds, but also social stability in the society and to the satisfaction of material and non-material needs of its individual members.

The ecology is a developing science and still more is the awareness of the value of the recent environmental crises that has stimulated the politicians, administrators and public in the understanding of natural ecological perspectives — ecology is having new look concerned with the system that supports man. Its emphasis is on the study of behavioural aspects implied with the interrelationship of the biotic and abiotic components.

2

Education in Human Ecology

Growing activities of human and his increasing demands on the biotic and abiotic components of environment has brought forth a change in his ecological philosophy. He is now directing his attention more on the study of habitat or ecological system as a whole than on their components. The biotic components are just a small past of habitat or ecological systems. In order to study the environment and micro changes brought in various components *i.e.*, water, air and soil, we study the climate and edifice elements. The distribution, adaptations and functional aspects of the organisms and communities are studied in greater details in ecology and relevant information is obtained from biological literature on biogeography systematic cytology, genetics and physiology.

It is, therefore, imperative for an ecologist to be acquainted with various disciplines of physical and biological sciences in order to get integrated picture of the environmental complex in relation to organic world.

Human ecology is the study of interaction between human beings and their environment. It is a means to improve people's life and environment, emphasising in-depth study of reciprocal effects that takes place physically, socially, economically and esthetically. Its basic approach is the conservation of natural resources both renewable and non-renewable and interlinks the science of biology as an interpretative science and elucidates the existence in the environment. It comes with the system that supports man.

Life on our planet originated much earlier than man came into existence. Mental set up of man has made him a super animal. Man is no longer slaving to the environment but through his technical skills he has to attain, he has to certain extend attained mastery over his environment. He creates favourable environment for his welfare in doing so, many a times, he forgets that he is a small component of nature and by disturbing nature and by disturbing the environmental balance, he is inviting trouble, change of forests, building roads and dams, industrial set ups and urbanisation, server and waste disposal etc., are some of the activities which are rapidly changing environments of our planet.

The paradigms of biodiversity conservation, which is, held by community whose survival and sustenance is linked to local biodiversity utilisation and conservation. For local indigenous communities, conserving biodiversity means to conserve the integrity of ecosystem and species, the right of those resources and knowledge, and their production systems based on biodiveristy. For them biodiversity has intrinsic value as well as high use value.

The utilization of biodiversity in people's economy is guided by a plurality of knowledge systems. The implementation of these properties, characteristics and use of this biodiversity is held with the local communities in different languages, and in their diverse epistemological frameworks.

Nature's diversity and diversity of knowledge system is however. Undergoing a major or process of destabilisation with the expansion of patents and intellectual property rights into the domain of biodiversity via the TRIPs Agreement of the World Trade Organisation. It is totally inadequate for the objectives of protection and conservation of biodiversity on one hand and protection of our indigenous intellectual heritage and social creativity on the other hand. Changes in our policy and law are being made with respect to our biological and intellectual heritage. The plant variety legislation is being introduced by the Environment Ministry.

No Ministry is addressing the unique problems associated with the objective of conservation of medicinal plant biodiversity and protection of indigenous systems of medicines and their practitioners especially in the context of TRIPs. The framework law for implementation of biodiversity conservation, therefore, can be covered in a legal framework, appropriate to the domain of biodiversity and indigenous knowledge for protecting our biological and intellectual heritage. In the absence of biodiversity legislation to perpetuate the piracy as intellectual property rights of western laboratories and corporation, will function as onesided legislation. Protecting our biological and intellectual heritage in the age of biopiracy requires the recognition and rejuvenation of our heritage and the evolution of legal system for the protection of this heritage in the context of emerging IPR regimes needed globally to prevent piracy.

To promote changes in the way the centre and states develop and employ human resources with the aim to make them and their sectors more productive and to modernize its agricultural sector at Nodal Centre for Human Ecology. To achieve these changes, policy and institutional reform would be carried out. The project would foster changes in the investment priorities resulting from policy and institutional reforms will focus on the process of : (a) improving the quality and relevance of higher education for community service through training programmes, (b) strengthening the capacity of participating states to develop and manage agricultural human resource.

To upgrade human resource management the project at state level would support development and management programmes (HRD & FIRM) including training focused on job oriented needs for : (i) Systematic need assessment; (ii) Training and trainers, (iii) Evaluation of training effectiveness; (iv) Better instructional facilities; (v) Improve management of state sector employees. This will redress environmental degradation, desirable resources and environment management practices, common property resource management, improve management skills, self-sustaining capacity, employment generation, mass awareness, income generation. The activities which are socially beneficial and economically viable like on farm, off farm and non farm will be undertaken for the benefit of the community by door step and on station technology transfer for functional literacy, through short-term training programmes to ensure continuous benefit to rural poor so that the resource and environment may be preserved for future generation.

EDUCATION AND TRENDS IN CURRICULA

Vocational/Career Oriented Programmes

Empowering people at different socio-economic and educational levels, which are appropriate to market, demand both in rural and urban areas. For students who have undergone general education courses in continuing education could offer specific skills, for those already in job to update skills or acquire qualifications for career advancement.

The school curricula of sixty's had elements of ecology and conservation of nature and its resources right from primary stage to biology courses of middle, secondary or senior secondary classes. The approach of integrating the teaching of science, social science even the mathematics and language with experience of children was extended to upper primary and secondary stage as part of general education. The environmental concept problems and introduction of social and environmental studies provided teachers with a curricular area in which the need and interest of children could be met through a focus on the local environment. Students must be given the opportunity to develop the sense of ownership, and empowerment so that they are fully invested in an environmental sense and motivated to become responsible active citizens.

Yeshpal Committee Report 1993 strongly recommended Gandhian approach to learning. Many other institutions in different parts of the country use different approaches and methodologies with same results. What runs through them all is the thread of spirituality and ecological thinking. All these institutions is still value education through course work as well as practices. It would be infect appropriate to describe the institutional framework for training and research in sociology, which has emerged after independence. Sociology in India is largely a discipline of borrowed concepts and methods derived from high prestige centres of learning in affluent wet especially in the USA and UK. This has resulted in chasing high prestige models into quick sand of pseudo-intellection. Sociologists have still not related themselves directly to the people and their problems and still reveling in counter productive — intellectualism. Our sociology does not address itself to the living concern of today and tomorrow. As a result it is not able to contribute meaningfully towards resolving many dilemmas of development.

The concept of human resource system therefore assumes that human being is great asset in the organisation. They are not merely necessarily evils to be reckoned with but they can contribute a great deal to the achievement of organisational as an asset with the unlimited potential in the concept of Human Resource Systems. The study of people at most is generally referred to as the study of organisational behaviour. It is the systematic study of the actions and attitude that people exhibit within organisation. The field of Organisational Behaviour (OB) is based on large number of systematically designed research studies. An organisation is a formal structure of planned coordination involving two or more people in order to achieve in common goal.

CONTRIBUTING DISCIPLINES

Organisational behaviour is applied behaviour science and as a result, is built upon. Contributions from several behaviours discipline. The predominant are psychology, sociology, social psychology, anthropology and political science. Psychology's contribution has been mainly at individual or micro level of analysis, whereas the latter disciplines have contributed to our understanding of macro concepts — group processes and organisational behaviour. The psychology explains or some times changes the behaviour of humans and other animals. The goals of human behaviour studies are concerned with developing people's skills, help to explain, predict and control human behaviour. The human behaviour studies after technologies that facilitate the control of people infect. Whether these technologies should be used in organisation becomes ethical question and should be aware, that the control objectives are frequently seen by managers as the most valuable contribution that human behaviour makes towards effectiveness on the jobs. Today's managers understand that the success of any effort at improving quality and productivity must include their employees. These employees will only be major force in changing out changes but increasingly will actively participate in planning other changes.

IMPROVING PEOPLE'S SKILL

The relevant theory and concepts that can help one, plain and predicts the behaviour of people. These are variety of ways to motivate people, how to be better communicator and how to create more effective team to be used in the job.

MOTIVATING THE WORKFORCE

Reconstituted into job requiring considerable higher technical skills or taking workers in other countries that will do the same work for a fraction of the wages. In this situation employee tends to perform either low skilled jobs or higher skilled jobs that means to maintain a middle class or upper class life style. This situation leads to a major challenge for managers. How to motivate who are making very low wage and have very little opportunity to significantly increase their pay either in current or through promotion.

SYSTEM DEVELOPMENT AND RESEARCH

Various systems for human resource systems need to be continuously designed, tested and reviewed. Data being collected also need to be analysed to give dues to the developing interventions. Development of human resource system can be conceived a system with interlinkages amongst its various components. The central concern of various components of HRS seems to be job. Job is defined in terms of critical attributes, background characteristics necessary like education and training and experience person has before he takes up the job we get training needs out of the critical attributes which also contribute to the concept of career planning of education and training. If we further expand we get into several new directions. The concept of human resource management is an approach to the management of the people. The behavioural science movement made two useful contributions to HRM.

Organisation Development movement. The Organisational Development (OD) movement advocated the implementation of programmes designed to improve the effectiveness with which organisation functions and responds to change, with particular emphasis on how people carry out their work and interest with one another. The total approach to organisational behaviour exerted a strong influence on many of these who in the late 70's and early 80's began concentrate on corporate culture as a central issue in the management of human resource. Different organisations have different traditions. In some organisations personal function may already include some aspects like general administration such as general administration such special features will certainly influence the design of some aspect of HRS. The traditional factor where HRS was designed as a part of the personal function. Where HRS was introduced, administration also put along with the department and designated human resource administration.

CHALLENGES AND OPPORTUNITIES

The human behaviour studies solutions or at least some meaningful insights towards solutions. The relevant concepts and theories that can help one, plan and predict. Each organisation has its own tradition. These influence the design of human, resource systems. If an organisation is highly centralised the designs will be own kind; if it is decentralised, the design would be quite different.

TRANSFER OF TECHNOLOGY

The process of transfer of technology cannot be examined in isolation. The effectiveness of TOT is greatly influenced by the linkages between and among those responsible for technology generation. The process of TOT could be examined wholly taking the process as major systems and the research extension client and input sub systems and its components for *Prioritisation of research needs and action points for functional literacy to deprives through training programmes and doorstep technology transfer* :

Empowerment of women for creation of an enabling environment where women can freely exercise their rights both within and outside home as equal partners :

— Generation of adequate human resource to meet the goal of promoting awareness, education through training and research programmes in postgraduate courses, post doctoral programme, short-term training courses for service and in-service graduate, doorstep technology transfer.

— Development of education, linked to social education from primary school age through adulthood of all groups of people.
— Enterprise resource planning — under university synergy programmes.

To know better the human resource potential available in the region in order to develop not only national HRD plans but also to develop sub-regional plans based on gaps analysis in supporting offered and to develop new mechanism for training.

Development of strategy for food production, income generation through genetic, biodiversity, natural resource management and education to improve the environment of the community on sustainable basis.

- Provide support with action by providing a consistent internal policy, goals and links.
- Provide leadership by taking part in linkage activities and giving them high visibility and priority.
- Encourage the group for commitments towards the farming related activities.
- Use mechanisms such as problem diagnoses, training and priority and set priorities.
- Promotion the farming community farmer participation as collaborators and evaluators of alternatives technologies.
- To college and circulate the information on training programmes to identify common training needs in the country.
- To develop such as ween sites to further develop its training activities in research organisations and management in the regions.
- Sharing the management experience whatever the management structure, a measure of autonomy is essential in order to maintain flexibility, objectivity and responsiveness to stakeholder, a single nodal centre handling research and education in agriculture together and able to coordinate effectively with other projects seen one model worthy of consideration. To ensure that research remains a profitable investment as well as to identify viable investment opportunities for research is

important. However, public sector funded research institutions need to be aware of their purpose in research for public benefit so that benefit from collaborating with the private sector should not be captured entirely to private sectors.

ADAPTATION, ABSORPTION

Adaptation is also the man's linking imported technology with domestic R&D. Adaptation therefore, is that it is consonant with the strengthening of the capacity of effective acquisition and mastery of foreign technology as well as building up of an effective research and development system — an essential element of attempts to foster technological self-reliance. Effective adaptation requires skilled manpower, which has atleast several years' experience in related production. The condition is not generally met in all but more advanced developing countries. The adaptation of technology however, is undertaken by engineers, technicians with an enterprise by industrial research institutions and by consulting engineers.

Absorption the effective absorption and diffusion is dependent upon an interest in applying the technology. The social carrier which has an interest in applying the technology. The social carrier could be an individual farmer who changes his pattern of production as a result of the introduction of improved plough. The carrier can be an institution. It might be an enterprise or agricultural cooperative that can develop, can promote the new machinery and agro-technologies. Absorption of technology in a narrow sense could be facilitated by policies, accompanied by adequate training facilities both in terms of number of persons trained and in extent of training. Free horizontal mobility has also be ensured to the pool of trained personnel.

TOWARDS UNIVERSAL EDUCATION

In the third world countries rural areas are the places where majority of people live, pursuing age old traditional occupation like agriculture/cultivation. To food the hungry million, the only way left out is to increase or multiply productivity with majority of the population access to education is a matter of utmost significance not only for the upliftment of women but also for general betterment of society. The inclusion of eradication of illiteracy, the universities have accepted in principal the important place of extension world, besides teaching and research with the participation of 83 universities and over 200 colleges, most of the universities and colleges have been involved in varying degree of programmes of eradication of illiteracy, population education and planning forums. Induction of other areas of extension such as rural development. Science for the people legal literacy, and environmental education has been undertaken. With the majority of population settled in rural countryside the development programme should aim at generating income and employment to one and all. Access to education is a matter of utmost significance not only for the upliftment of women but also for the general betterment of society. Incidentally some of the programmes with accent on women's participation have indirectly helped women to earn while indoors without distributing their normal households chores.

The development trends have picked up from macro level perspective. The filtration frames the wider Indian experiences in rural development wide range of theoretical inter-relations already settled in growth theory. With the formulation of UGC policy frame 1977 a process of involving university system as such to participate extension programmes was initiated with the specific objective of securing equal importance of extension as a third dimension with teaching and research. Though some universities were conducting continuing education programmes and universities involved in substantial way in the National Adult Education programme of Government of India. However, the infrastructure is too inadequate even for minimum performance for institutionalising need based extension programmes to involve identified communities in the life long learning process yet not operationalised in most cases.

Institutionalisation of extension needs academic acceptance which is automatically ensured when the activities enter at a stage of studies and research which enlarges our understanding and contributes to the knowledge. Extension activities can enter into research through project work or field based dissertations at undergraduate and postgraduate level.

GLOBAL PROGRAMME, MANAGEMENT

Urbanisation — Basic Services and Community Participation

The rapid pace of urbanisation in the developing countries and the consequent pressure on urban civic amenities matters of serious concern the world over. Experts, scholars and administrators grapple with the problem for all urban poor access to adequate basic civic amenities in burgeoning cities of developing countries. An important element is local community particularly empowerment of urban communities. Increasing urbanisation accompanied by increasing concentration of the urban population is bigger cities has not only posed the problem of pollution, crime concentration of the mushrooming of huge metropolises congested housing but has housing led to severe pressure on the basic civic amenities and consequently a decline in the living standard of the city dwellers. As a result although the situation is no better in rural areas, the urban areas suffer from inadequate provision of safe drinking water and sanitation facilities. The problems belonging to the developing world are expected to become acute. It may even lead to international welfare. Many factors such as resource crunch, lack of proper management, too much dependence on central authorities as well as very little involvement on the part of the community of the specific problems of patterns of urban growth. To learn form the experiences of other countries suggest concrete recommendations to the situation. In addition it aimed at assimilating further thinking with the status of basic services in terms of drinking water and sanitation facilities, problems of decentralised planning community development and the role of socio-action group. The dimensions the problems of decentralised, planning community development and the role of social groups.

Global warming, diminishing vain forests, protection of ozone layer and preserving genetic biodiversity are all subjects of discussion today but closer to ground is the problem of water pollution we have recognised that how much water is needed but rather how best the water demand can be managed and how to optimize the use of available water.

Though only seven per cent of world water is used for domestic demands and 70% for irrigation, the efficiency of water used in agriculture in most countries is around 30% only. The inefficiency is no less in urban areas. Leakage and unaccounted for water in many cities and many countries is around 30% only. The efficiency is no less in urban areas. Distribution arrangements and existing tariff systems have hardly any relevance for water conservation and wastage reduction. The contamination of water resource is now widespread in many developing countries brought about by industrialisation as well as untreated sewage. Surface water pollution is also major problem. Unfortunately, the cost of conventional sewerage system continue to be very high and the level of technology development and application efforts limited, while man has succeeded in putting one of his species on the moon, he has not been able to deal with his own wastage in an efficient cost effective manner. Technologies like an aerobic treatment and those of recovery of resources from sewage continue to be applied on the limited scale only. The Ganga action plan launched in this country in 1985 was the first comprehensive plan of action to address in an integrated manner, conveyance and treatment of sewage, resource recovery through methane gas, fish farming and irrigation, sustaining aquatic diversity and river quality monitoring and management. But the plan is limited only to Ganga.

Management

Management is concerned with the optimum utilisation of various resources for the attainment of organisational goals and objectives. The proper management of human resources plays an important role for attaining such objectives. The behaviour science is a wholly distinct discipline. It deals with the behavioural aspects of human being, which analyse the factors, which influence human behaviour. Every manner by virtue of his own experience as a person, the managerial job involves all kinds of people problems, but he might also argue that he knows from his own experience how do deal with most of those problems. Behavioural scientist has worked on human problems in organisations. But the nature of the problems studied has evolved from individually oriented industrial psychology focused on problems of recruitment, selection and testing to organisational psychology and sociology focused on managerial, style. How groups and teams work, inter group collaboration and competition, and total systems analyses of organisational and managerial effectiveness.

Management is the creation and control of technological and human environment of an organisation in which human skill and capacities of individuals and groups find full scope for their effective use in order to accomplish the objectives for which an enterprise has been set up. It is involved in the relationships of the individual, group, the organisation and environment.

The role of urban local bodies in providing water supply has verified from country to country. A community plan to demonstrate to upgrade and extend to such areas to meet further needs. Extension of social surveys and distinct packages for the demonstration project to find the entire project on a grant basis. The only effective way to ensure that cultural factors are incorporated into service programmes would appear to be by involving communities in the design of the project. Water consumption and waste disposal impinge on extremely personal attitude towards bodily hygiene, which are largely cultural in nature.

To conduct need based study, survey, adaptive research and development field trials, demonstration, dissemination of appropriate technologies among rural poor. Strengthening the existing institutions and setting up of new institution to test technologies suitable for specific geographical area making modification where necessary and initiate the process of transfer of viable technology through demonstrations, training and manufacture. The new institution will act as a focal point of referral of technology gaps in the area and generation of matching solutions to answer the technology needs, identified by government and non-government institutions. It will also conduct training programmes at various levels, especially trainer's training programmes through the demonstrative achievements in the adaptive research and development and dissemination of innovative rural technologies for five years.

The scale of urban growth in many developing countries during the past four decades has been without precedent in human history. Although providing storm drains are separate for servers, open channels with earth banks are generally appropriate in that they are inexpensive to construct and easier to maintain.

Government can work directly towards specific community objectives, or by offering contracts that serve those purpose. Furthermore traditional engineering approach often encourages expansion of infrastructure and supply augmentation rather than alternative such as demand management. Still, investment in improved technologies, such as design of sewers, design and construction of main holes, inverted siphons, protection of water supply, waste water facilities for sewage keeping with current usage for more conventional municipal waste water collection and treatment systems. Innovative approaches to collection and treatment particularly for the very small municipal systems practices for safeguarding of the public health or protection of water quality, for desirable procedure or methods with deviations subject to individual consideration. The wastewater treatment works for safety, design, essential facilities and laboratory. Settling sludge and scum removal, protective and service facilities, Sludge processing, storage and disposal, Biological treatment systems, Disinfection — chlorine disinfection, dechlorination, ozone. Supplement treatment — handling and treatment of septage at a waste water treatment plant.

Innovative construction of a municipal garbage composting facilities, with a small-scale energy efficient sewage treatment plants and fish to change sewage into clear water. Any technology that encourages the recycling of resources should be seriously considered by communities that attempt to plan sustainability. The expenditure may be taken as solution to many policy problems.

Status of Water Supply and Sanitation

The global consultation on safe water and sanitation which met here in Delhi in 1990. A part of problem in measuring coverage is that there are limited to simply arithmetics, usually most national reports desired the total population of a whole town as covered if there is some kind of water supply are calculated even more simplistically by dividing the total installation capacity by population. We need to go very far for the truth.

EVOLUTION OF RESEARCH SYSTEM

Expanding the research for pilot scale testing of the technologies already developed for commercialisation would be valuable to organisations within and outside UGC system. This fund support would also help to broaden UGC base by drawing upon the strength of individual research institutions, which in term can fill gaps in unforeseen but critical areas of research related to regional problems besides contributing to developing staff skills through extension outreach and study tour — entry of institutions outside the ICAR/SAUS systems for enhanced inter-change with national and international centres of scientific innovations.

Several changes have taken place in the organisation and functional set up of Institute of Home Economics and in their research and development programmes. Commensurate in these, there has been no proper emphasis towards development of management skills of faculty members to make national research programmes more efficient and effective. Project for R & D activities are time bound, and lay high priority to meet changes such as poverty alleviation, food and nutrition security, natural resources conservation and overall sustainable development of agriculture and human resource; efficient managers are required.

Selection of species — Energy farms are still in the conceptual stage. It require detailed studies in genetic, agronomic and other aspects particularly the selection of appropriate sites and species, spacing and fertilizer usage and resistance to pest and diseases to maximise dry matter production and increase availability of alternative source of energy. Following criteria need to be adopted for selection of species, case of establishment and regeneration after harvesting, freedom from pests and diseases, ability to reduce transpiration losses in arid areas and minimum drain for nutritionist. Unlike aquatic plants, there is need to replace each harvest with a new crop as quickly as possible, the plants that have high sprouting potential following coppice cutting are preferable to those requiring replanting, using or breeding plants with leaves whose shape and orientation makes maximum of sunlight is another promising area to increase biomass production (i) ability to conserve water, (ii) minimum drain of nutrient and ability to improve soil quality. The enormous diversity in the range of ecological conditions in India offers problems in afforestation and biomass production. For successful plantation trees must be raised in the nurseries later transplanted in the field. In order to achieve success in this newly developing challenging areas, steps need to intensify on Resource survey and technological planning.

The Study of Globalization

To work with the Centres of JNU for organising pre-service and in-service courses, refreshers training with the centres of excellence like Zakir Husain, Centre of Educational Studies, School of Social Sciences, School of Life Sciences, School of Physical Sciences, School of Environmental Sciences and Centre for Biotechnology.

For action to be taken in order to overcome institutional bottlenecks to participatory programmes.

- For the sustained process of endogenous innovation generating activities as indicator of success for continuity of learning in academic structure for the involvement of curriculum work and re-animating the collaborative process of generating knowledge in the classroom and taking independent research by the students.

Basic concepts — Paradigm in shifts in human services that are promoting change for better outcomes for children, families, neighbourhoods and communities processes and tools — ways of operating that can facilitate the new way of doing business.

Best practice — models — Tried and tested human service delivery basic research and evaluation reports and field experience the best information currently available than decision makers can use in human service programme development.

Technology assessment and refinement through institute village linked programmes require linear mechanism before their transfer. The programme approach involving various technology components relevant to the armours in varying agro-ecological zones for a perceptible change.

Discoveries of development of new technology by scientists usually in the form of principles, the bits and pieces of which are to be brought together for the development of appropriate technologies for the farmers. However, commercial farm production systems with emphasis on off farm and on farm value additions? The entire programme seeks to develop multiple options for different target groups through a participatory mode.

Regional and global state of the Environment reports traditionally do not systematically document existing policy responses, let alone assess their impact and effectiveness or analyse factors contributing to their success or failure. This trend is changing, however, as national and sectoral SOE reports pay more attention to the policy side of the environmental situation. Improving our ability to produce overview and assessments of societal responses to current and emerging environment issues is one of the aims of the GEO Report Series.

Effective policy setting for sustainable development requires a blend of policy instruments that addresses the social fabric of life, ensures effective institutional arrangements, improves the economy and protects the environment.

Global Frameworks and Conventions and National Policy Initiatives

Since the 1972 Stockholm Conference, the number and scope of international environmental policy responses have increased significantly. Initially, the established global strategic planning framework usually has a sectoral nature. Examples of this include the United Nations Plan of Action, the World Conservation Strategy (1980), the World Soil Policy (1982), and the Tropical Forest Action Plan (1984). Under these global programmatic framework, national action plans strategies and policies were prepared.

At the same time, multilateral agreements and conventions were negotiated, resulting in some 200 legal instruments that target regional and global action to protect the environment. Some of the new, currently negotiated legal instruments and action programmes. Several of the global conventions call for national inventories action programmes and reporting mechanisms, such as climate change country studies, biodiversity country studies on sources.

Rules and regulations to implement these international agreements at the national level. The main policy instruments used by governments to translate international agreements into concrete action are based on command and control principles.

Although initial co-ordination of such national level efforts may not have been very good, today national implementation of global programmes is better co-ordinate and often manifested in single overall environment action plans. The nature of national instruments has changed from a narrowly defined sectoral focus to a more comprehensive and anticipatory approach to protecting ecosystems. Increasingly, socio-economic factors are taken into consideration. Many countries have established SOE reporting programmes, which attempt to report on the status and trends of various sectors in an integrated way. In addition, national legislation has been developed at a rapid pace in the past decades. Governments of many countries, for example, undertook significant amendments to existing natural resources legislation in order to deal with environmental problems arising from over exploitation by commercial and industrial sector. It is the emergency of resource-oriented legislation and ultimately, system oriented and integrated legal regimes aimed at long-term management and more sustainable use of natural resources. In industrial countries, some of these already existed before Stockholm; Wildlife legislation for instance, gradually incorporated the concept of maintenance of a safe minimum stock through protection of vulnerable species.

Changing Environmental Policy Perceptions and Concepts

Many parts of the industrial world achieved economic growth and wealth in the past by having unlimited access to environmental resources and little thought was given to the impacts of growth on the environment. They then sought to improve and protect their quality of life from the environmental impacts of their development — initially from the effects of pollution on air and water and more recently from hazardous waste, biodiversity loss and global climate change.

Significant improvements in environmental protection have been made in many parts of the world through such strategies and action. Yet the state of the ecosphere as a whole continues to deteriorate. Developments in the different regions and explorations of the future show that an increasing number of environmental constraints will be encountered in the next few decades. Although command and control standards are effective in many cases in terms of short-term environmental improvements, the costs of implementation, enforcement and compliance are high and may hinder long-term economic development. Furthermore, although such policies have proved to be efficient for pollution control, they are less effective for problems associated with the management, protection and conservation of natural resources, particularly when a large number of different groups and people use these resources.

Effective policy setting in support of sustainable development indeed requires a more diverse mix of policies that address the social fabric of life, ensure effective institutional arrangements, improve the economy in all its sectors, protect the environment, and so on. Consequently, countries are complementing command and control policies with, among other things; market based economic instrument methods to achieve cleaner and more resource efficient production systems, and efforts to change consumer attitudes.

A 1996 UNEP study explored the question of the environmental effects of stabilisation and structural adjustment programmes. Under standby the answer depended on a large number of interrelated factors, implying that economic policy makers cannot apply a simple, standardised set of reforms to any given economy and expect predictable, consistent, or even beneficial results. Macro-economic reforms may even have unpredictable and mixed effects, depending on the situation in a country or threaten economic growth and environmental integrity for the future in certain cases.

The share of people enjoying medium human development rose from 55% in 1975 to 66% in 1977 and the share is low human development fill from 20 to 10%. But these trends mask great unevenness in the advances and in the new set backs. Despite of tremendous progress in the 20th century the world today faces huge backlogs of deprivation and in equality that leave huge disparity within countries and regions. Poverty is every where. Measured by the human poverty index more than a quarter of the 4.5 billion people in developing countries still do not have some of life most basic choices — survival beyond age 40, access to knowledge and minimum private and public services.

- Nearly 1.3 billion people do not have access to clear water.
- One in seven children of primary school age do not have access to clear water.
- One in seven children of primary school age is out of school.
- About 840 millions are malnourished and an estimated 1.3 billion people live on income of less than $ 1 a day.

Human Development to Make Growth work

Social welfare is as old as humanity itself. Helping the needy and the person in distress has been in existence particularly in all the civilisation communities from time immured. Social welfare has recently developed into a scientific programme poverty, sickness suffering, social degradation have always been known to man in long course of history of mankind, but after the industrial revolution new social problems were erased, which could not be faced by the older human institutions — family, neighbourhood and local community. In fact this was not the situation only in industrial societies particularly in the west, but such a situation is being created in the developing countries also. Apart from the humanitarian aspect, the development of scientific knowledge provided the tools for investigating the causes of poverty, of human deficiencies and dissatisfaction with the sole object to find out a remedy for various social problems facing the society. Hence social welfare development into a scientific concept based on a scientific knowledge of the problems facing the society, particularly in free society with the object of finding out ways and means of solving these problems.

Thus understanding of social relationship, social psychology and social economics has helped to understand the social problems and the ways and means of organising social welfare services on scientific basis with the help of a tested body of knowledge. An orphan child needs more than food and clothing in children's institution. The problems of aged and inform persons are partly social and partly psychological in nature. Therefore, the institutions of this basic human needs. There is a need for social welfare programmes like balwadis, holiday homes, youth hostels etc. which help in the prevention of occurrence of social problems. The programmes of social assistance have also to play a very important part in building up the human material. Various services like creche, orphanages, and homes for the blind, workshops of the orthopedically handicapped persons, community welfare contracts were set up by the interested citizens for the purpose of providing welfare services to the needy and the destitute.

Whereas, there has been a large measure of goodwill and dedication, there has been dearth of technical knowledge required to run various welfare services. Foreign literature was available on various social welfare services but this was not suited to the requirements of Indian condition. Attempts have been made in certain areas of welfare, work to provide material on different kinds of social welfare service, like minimum standards on child care institutions, worked out by the Indian conference of social work, and the minimum standards for different welfare programmes by the central social welfare board etc. Some useful material is also available from the reports of various committees and proceedings of conference and seminars on social legislation. All this material is not only inadequate but is not available to the social workers on one place.

The traditions of social services were influenced by various social and economic factors. The social institutions like the joint family, the caste system and the village panchayats provided mechanism, which would meet the needs of the sick, the aged, the destitute and the handicapped members of the community. Industrialisation and urbanisation also changed the pattern of life. New social problems cropped up, for the solution of which the existing social institutions found themselves ineffective. Without the development of social science it became possible to make a systematic study if the social problems in order to solve these on a comprehensive and scientific basis. We, therefore, started moving is non-being laid on the application of scientific methods to the solution of social problems *i.e.* by studying the social and economic factors of the problem and acquiring special skills and techniques, in dealing with them.

Social welfare board. Social welfare services were brought on the political map of India 1953 with the setting up of the central social welfare as a part of First Five Year Plan. During the recent years there has been increasing realisation that with a view to improving and preserving the human capital which is one of the most important factors of production of any economic development. It is necessary to provide opportunities for growth and development of the child. The social welfare field has so far been a combination of various social welfare activities without aiming at an integrated approach either to the child or to a family or to the community.

The concept of service of one's fellow men has been an integral part of Indian traditions. The ruptures shared the responsibilities of meeting the needs of the individuals and groups in difficulties, the rich and the members of the general community shared the individuals and the group's difficulty. Religious institutions also played a very important part in providing social service. We come across men and women and group of persons who devote their lives to the service of helpless. The measures of goodwill and some leisure time with the industrialisation and urbanisation, our social problems are becoming more and more complex. There is also a growing recognition that for handling such complex social problems of social, economic and human relationship, we require certain technical knowledge concerning human behaviour, social institutions etc.

Child welfare. In order to preserve and to improve human capital, investment in the child is very important, in any plan of economic development. The child should be looked after as much as look after our dams, factories etc. and that in many welfare programmes priority should be given for meeting the basic needs of child. These needs are healthful food, adequate clothing, comfortable home with love and understanding, educational facilities, healthy services recreational activities etc.

In order to satisfy these needs, the child should be looked after as much as we look after our dams etc. and that in any welfare programmes priority should provided all basic services necessary for its physical, mental, emotional and intellectual growth. As investment in basic preventive services for the child is a must in order to prevent blindness, deafness, mental illness or other social handicaps. This will minimise expenditure on the services meant for the treatment of these social ills. These services must be coordinated and integrated so that the total needs of children are met and we have a complete child physically, mentally, and intellectually strong. It may not be possible at this stage to provide all these services to the entire child population in age group of 0-16 but it is necessary to make start somewhere. The government of India worked on this principle and have initiated demonstration projects of integrated child welfare services in similar community so that all the needs of child are met on integrated comprehensive basis, methods adopted and the results achieved are demonstrated to the community so that the programmes could spread in future.

Child welfare worker. In order to meet the total needs of child and to make the programme of integrated child welfare service is successful it is necessary that there should be technically trained workers at the field level who have understanding of basic needs of the child and the knowledge of concept of integrated child welfare services. The Indian council for child welfare has been working on the schemes to provide special centres of child welfare workers 'Child Ball Seika'. The Ministry of Education and the Central Social Welfare Board approved the chime. Training centres have been set up in different places in the country. Each centre would train a batch of 5 women workers for a period of two years who would be give special training in handling children in the age group of 0-6 years and meeting their needs in education, health, nutrition and welfare. The training would thus be multidisciplined in its approach so that worker is able to asset in the provision of basic and integrated services to children.

Child Welfare Service

The most common form of service to children so far has been institutionalisation for an orphan, destitute, blind not very healthy because institutionalisation losses personal touch and lacks homely atmospheres, necessary of the growth and development of the child. Infants under 3 years of age who need warmth, love and security normally children between 3 to 6 years who have special emotional needs, orphan children who need special individual care cannot be benefited by group care. While providing services to the handicapped childcare should be taken to see those resources of both family and community are utilized as far as possible. The services of children in the context of present welfare programmes in the country may however be classified to both normal and abnormal children.

Absence of Family Life and Lack of Parents

Affection is Responsible

Affection is responsible for insecurity and emotional disturbance. To some extent disciplinary problems in institutions can also be reduced. Group recreation gives a child, a sense of security and enables him to make some adjustments through the community life. For good ovation through recreation cast barriers and group prejudices can be broken. The following are different forms of recreation :

Physical : sports, athletic, souting

Creative : Handicraft, music, dramatics etc.

Mental : Discussions, debates etc.

Social : Picnic, trips, holiday home etc.

Various recreational services today are : play centres, hobby club, holiday homes, etc. The principles of recreation include consideration of cultural and economic factors of community life ensuring participation by large number, focusing attention on the group, taking into consideration the needs of different individuals based on age, interest, abilities, personalities etc.

School health and social service the child has to move out from this primary group and get an opportunity of associating himself with institutions like the neighbourhood, the playground and the school. Therefore it is very necessary that proper conditions be created for a child to adjust to his school environment and make use of educational and recreational programmes for his emotional and intellectual growth. A school particularly in India, concentrates only on the training of the mind, it is essential to stress that the whole child who is in the school.

Child guidance clinic. In order to avoid maladjustments and behaviour problems, services of a child guidance clench are very essential without the help of trained psychiatrist and psychologists these maladjustments may be checked at the earlier stage of their development. The clinic serves children with behaviour problems and nervous conditions. These problems are indication of some inner disturbances it may be seen that the work of the clinic is concerned with children with unsatisfied emotional needs and not only with physical and mental difficulties. In India, because of various reasons it may be possible to focus on mental ill health and the need for treatment.

Physiological principles, methods and techniques are difficult to formulate. They lack precision of the other sciences as they deal with human beings that are complex and are always changing. It needs the services of a trained and qualified psychiatrist and other members of the team. Apart from professional training he should have the knowledge of local socio-economic psychologists a cultural factors child guidance is thus a very useful service. Handicapped child — physically handicapped children the bluing, deaf, dumb, crippled etc. There are two main aspects of the services for the handicapped children.

Prevention of handicapped. By providing minimum basic services to the child, so that they could become useful citizens by providing education training and rehabilitation. We require specially trained workers for the education and training of the handicapped child.

Mentally retarded child. Mentally retarded child is a special group. There are educable mentally, trainable mentally retarded and the totally dependent mentally retarded children.

Apart from assistance in feeding, clothing etc. it requires protection from danger and he cannot service without outside help throughout his life. This field of child welfare is broad. Therefore, child welfare has got both preventive and treatment aspects.

Social service by youth. Youth has always stood for energy enthusiasm. If their energy and enthusiasm are not properly harnessed to constructive ends, the misguided youth may use these from anti-social activities.

In order to harness the energies towards social service, a suitable organisation has to be thought out and built up. There is need to have strong and efficient organisation to give necessary guidance to the youth, so that welfare services in the community are provided under the direction of competent youth leaders. There can be three types of organisations.

1. Agencies working in thè field of rural development and running community welfare centres, adult literacy, relief workers.
2. NSS of colleges.
3. An organisation of youth under 'Yuvak Seva Dal'.

Area of work. The work to be done by youth may be of two types : (a) work of permanent nature and (2) work of temporary nature.

The work should have some continuity sporadic and half-hearted activities in any community are of no use. Students should contribute to work comprehensive programme which should be run by an established agency of social service league of a school or college. The gap in work has to be provided by someone, because the students are unable to devote full time to social service. This gap could be filled by whole time workers of the agency or by teachers in case of college social service league.

The youth intend to help should have an understanding of the rural problems and the rural people. Some times a student may attach himself to an existing organisation engaged in rural welfare or a group of students might even set up all organisation of the students who could adopt a nearby village for organising welfare services.

The youth working for rural reconstruction have to bear in mind certain principles of working with rural community. The fundamental principles are that of self-help. Any effort to do some work for the community rather than with the community will create a feeling of dependence on the outside agency instead of any result.

Great caution should be exercised in approaching the rural people. It is desirable to first approach the village leader and talk to them about the purpose of party visit to the village and then discuss the village problems with them and see how far they are prepared to do something about these problems. In many cases, it may be necessary to contact some outside agency or Government authority, in order to get government and outside support. The students could act as the interpreters or spokesmen of the community needs in such matters. The student should start with younger generation, say children by organising recreation for them and then go on to other groups and different areas of work.

Nothing should be imposed on the community efforts should be made to improve their life without imposing an urban pattern of life on them. They will become antagonised, if we try to judge them from our standards. We have to start at the level of the community. We have to respect and take into consideration their superstitions and customs etc. The worker should try to win the confidence of the village people, otherwise it may not be possible to get their co-operation and support. The students should be given some kind of orientation before they go to work in a village.

Programme. The following are some of the programmes in which youth and the students can associate themselves with :

1. Cleanliness, sanitation and health;
2. Campaign against disease and epidemics;
3. Adult education;
4. Cultural activities and recreation;
5. Village industries;
6. Improvement of educational activities;
7. Organisation of village youth;
8. Civil defence work;
9. Community work during emergency.

HOME SCIENCE CURRICULUM IN HIGHER EDUCATION

Education determines the level of aspirations, application of technology productivity, sensitivity and efficiency of education. These constitute the factors basic to the process of development. Higher education helps people to bring about desired changes by utilising constructively the knowledge skills and conditions made available by scientific technological advancement. Education contributes in a large measures, not only to the elimination of the ideas that women are inferior but also helps to wipe out the inferiority of their status in the family and community. Home Science education for living is a synthesis of arts and science and the fine arts, interwoven together harmoniously to foster, aid and nurture the physical social and spiritual development of individuals family and community. After a long period of struggle for survival, Home Science today enjoys an independent status in educational institutions right from pre-school to the university level and in the curricula of extension training, continuing and adult education and non-formal education systems.

Changes in curriculum takes place through both innovation and development. They are caused essentially by two forces :

- Internal emphasis of the faculty who by process of creating new knowledge influence the university to change the curriculum.
- The external impulses, that is the circumstances outside the university, which force changes in the curriculum of the university to meet the emerging demand of the society.

The goals of the curriculum are derived from the social realities and its tasks and functions are aimed at changing these realities while the professional values undergrid and serve to place goals in the desired directions. The organising elements of such a frame include the value of profession, the skills, and the methods that are developed for the professional task and the major themes and concepts in teaching research and extension.

THE GRADUATE SCHOOL

For the involvement of learning community a postgraduate community which provides its members with support and stimulation. Academic staff facilities for learning in variety of ways — lectures, group work, individual research and tutorials. To help the student to become self-sufficient in organising own study and development. Academic staff involved in postgraduate work is active in research and constancy, producing publications regularly. Research is disseminated in teaching at social policy research centre. Postgraduate and post experience programme to focus on :

- Criminology
- Social policy
- Psychology
- Geography and environmental management
- Social science research methods
- Psycho analysis
- Social work
- Research degree.

Collaborative Partner

Collaboration is a key aspect of an approach to research and current partners include facilities and quality. The commitment to quality as well as continuous improvement to the programme.

Following programmes will be added to those currently available :

- M.A.M.Sc. countryside management and other land based profession
- M.A. in Indian Criminology
- M.A. Social Work Studies
- M.A. Development in Practice.

Studies for a taught Masters Degree in School of Social Science :

Flexibility builds into postgraduate programmes in two ways : M.A. multidisciplinary programme and through the choice policy to take option from criminology. Option may be chosen from postgraduate programmes in other schools in the university *e.g.* M.A. Humanities.

THE TEACHING AND RESEARCH PROGRAMME AT SCHOOL OF SOCIAL SCIENCES, JNU

The teaching and research programme in the school have certain innovative elements while ensuring rigorous discipline oriented training in a centre. Interest in generated in multidisciplinary study and research, by encouraging students to offer courses in other centres depending upon the aptitude of the student as well as the relevance of these courses to the main discipline. Many students take advantage of this opportunity particularly at M.Phil. level and research activity consequently has been enriched. Many staff in the administrative offices, the library, computing the other support services contribute to making the programme enjoyable and worthwhile experience. Work experience internship modules which can be the place of one or two taught options. Students are assigned in funding higher quality work experience placement in a range of organisations. To equip the student community with the practical skills required in todays employment market.

Internationalisation of Home Science Education in Health Sector

Under the influence of US pattern and with the expertise available the curriculum tinder went a number of rat changes subsequently not only the undergraduate curriculum in various colleges were started but also only for postgraduate courses in various disciplines. Job oriented programmes were introduced by the University of Chenai at B.Sc. level like B.Sc., Nutrition and Dietetics, B.Sc. Food Preservation, B.Sc. Child Family and Social Welfare and B.Sc. Design and Environment.

HOME SCIENCE AND EXTENSION WORK

In view of the emerging needs, it is necessary to prepare the future Home Science graduates with an aptitude for and skills with working with rural families in the capacities of extension workers, trainers for the extension functionaries and extension supervisors. Ministry of Food and Agriculture of Govt. of India provided grants to few selected urban home science colleges for establishment of extension departments to develop the concept of extension higher level of Home Science education, research in the fields of relevant to rural families. Today Home Science extension departments are monitoring the outreach programmes of the colleges/universities.

Home science and women studies have emerged as a new discipline study higher education. In the context of health sector is increasingly realised that one of the most important reasons for sub-optional performance of health care institutions in poor quality and in appropriateness of education and training of health care providers, resulting lack of problem-solving competencies and skills.

The demographic transition, changing life styles, information technology institutions and increasing emphasis on globalization are further adding complexities in the task of education and training.

A great deal of work on foods and nutrition has been carried out globally over the last two decades. The current scenario of poor nutrition profile is partly due to minimal involvement of trained nutritionists with home science background in nutrition and other welfare activities. The traditional textbooks of published literature on food science and nutrition is mostly based on western experience and is not properly oriented to the current needs of the country.

Vocational home science education is most appropriate for vocationalisation as it is a unique combination of many sciences catering to man's health, human relationships, effective management of resources, upbringing of children, feeding families, tailoring their own garments, housing, caring for the sick and application of art principles in daily living. At present vocational courses are confined mainly to technical training institutions and polytechnics. The existing course, which is now at this level for the training of teachers, para medical personal agriculture and the women's polytechnic could come under Home Science programmes.

Developing the curriculum — against this back drop. It is necessary to examine and re-examine our courses in home science as they exist today and make their relevant to the needs of students. Community and nation.

COMMUNITY EMPOWERMENT AND SOCIAL JUSTICE

Empowering people at different socio-economic and educational level with marketable vocational skills, which are appropriate to market, demands both in rural and urban areas. For students who have undergone general education courses in continuing education could offer special for those already in jobs to update the skills or acquire qualification for career advancement. Need based relevant education programmes to fill in the gaps in their intellectual growth, enhance professional and technical competence and promote undertaking of contemporary issues.

Stimulating the other departments of the university to undertake similar activities in assessing the need of the community with respect to their discipline for continuing education of graduates extension education and field outreach. Developing curriculum and multidisciplinary programmes involving faculty, students, from various disciplines. This can help in developing the discipline through academic programme and courses in criminology, social policy, psychology, geography environment management, social work and research degrees committed to high quality learning and research.

University Grants Commission plans to be major national partner in the promotion of human rights, culture and considers it to be one of the major objectives of higher education. However it is essential to consider the specific national environment of the user and to adapt training material to local situation to ensure their suitability in different context : a national legislation and policy on women, the social status and juridical practices relation to women workers rights, the discriminatory aspects in gender relations at national level; and the existing level of legal literacy among women workers, women organisation and community at large are some of the important issues earmarked for the empowerment of women.

Education for Human Resource Development

The human resource development movement, began in the country during the 80s of the current century when Government of India constituted the Ministry of Human Resource Development, it was realised for the first time, both at the macro and micro operations that the development of the country organisation dependent to a greater extent on the development of human resource of different types rather than the development of material resource such as land and building, capital, machine, material, technology etc. The HRD movement has taken off in the country. Education are mostly legal or economic not developmental. Cabinet Secretariat approved the department of women and child development national policy for the empowerment of women. A new scheme on integrated women's empowerment was approved. For the holistic development of the child department of women and child has been implementing the integrated child development the department continued to recognise the contributions and potential of rich and vibrant civil society organisations, especially the women's group, all over the country and forged a strong partnership with them for the implementation of many of its innovative schemes and programmes or the empowerment of women and development of the child.

Women Economic and Social Empowerment

More scientific inputs can really go into improving the traditional system and sources. Small economic development programmes which promote greater support by village men and local institutions to women's lives in order to make them more active and efficient in their participation in economic and social development particularly in the area such as maternity protection, child care facilities, nutrition and health care both preventive and educative literacy and non formal education. Building up of linkages with appropriate agencies, training of staff, implementation of experimental pilot projects in selected districts for effective propagation of appropriate technology, financial assistance for economic development and support to S&T programmes for rural and tribal women. Training should be imparted to develop skills and competence of staff.

Research is necessary input for successful implementation of any programme to be undertaken by rural/tribal women organisations. Exchange of ideas between functionaries of the schemes. There is need to develop and implement sizable number of centrally sponsored pilot programmes in each district to develop integrated projects for tribal women with a multidisciplinary and multisectoral character. The main thrust on the structure of such a project would lie in providing integrated service with minimum input in economic development both in agriculture and industry, education programmes particularly through adult education schemes health services particularly better maternity care, and child health, nutrition, family planning, scientific culture outlook can act as primary weapon to fight discrimination against women. The project should create the necessary environment or institutional arrangements to take

S&T to the doorstep of rural families, in view of the absence of transport facilities.

In order to unlock the human resources to the mainstream of national life, their present social status is to be improved, which has deteriorated under the impact of modernisation and industrialisation with especial emphasis on adult education for women through mass communication media. The communication technologies should be well developed on operational management of environmental projects with participation, advocacy skills in the areas of environment knowledge of legal provisions, land resource management, biodiversity sustainable agriculture and promotion of linkages between groups and government banks, panchayats, research institutions, in the area of environment development and gender issues in operational management like knowledge of legal provisions, policies in the area of micro enterprises. Such awareness only will help them fight against exploitation.

Appropriate Home Science of developing Asia

Asian Regional Association for Home Economics and International federation of Home Economics in its one-day programme was hosted by the faculty of Home Science M.S. University on Jan. 1992. Following suggestions emerged out of the deliberations that there is a greater need for the interaction and information and greater international exposure for academics by teacher exchange programmes. Greater support is urged from UGC for establishing new Home Science Colleges accessible for rural students. Further greater financial assistance is stressed for updating and strengthening the existing college programmes.

Home Science in School Programmes — to introduce home science education at plus two *i.e.* at Higher Secondary School level and the curriculum be erratically planned taking into consideration the regional requirements leading to quality programmes at higher educational level. Home Science institutions are made coordinating agencies between related government departments; institutions and voluntary organisations to provide needed human resource development component. The extension programmes should be made part of B.Sc. programme for greater exposure and interaction with community. In the view of fast pace of change it concerns curriculum and its structure; community outreach; Research comparative studies; clientele.

The curriculum and its structure that develops vocational or professional competencies in teaching, communication and extension that develop awareness of available employment opportunities. The community out reaches formal and non-formal education of urban slums, and extension input for rural people and facility and university. Appropriate research on issue relation to women stratified according to employment, education community either than only urban and rural women.

Local, block and state level surveying, samples and applied action should co-relate designs rather than fundamental research only with conceptualised sample of women alone. Both men and women and family or an integrated approach should be considered. To meet the need of new slot, training programmes should be designed and implemented by the extension education departments in collaboration with relevant departments and agencies through short-term courses. Specific training for vocational at various levels high to low for example.

Dean-Professor-Teacher-School-Teacher-Lab Teacher to create acceptance of education and extension graduates for new areas of jobs. Include self-employment in extension education and ongoing continues interactions between education and employment agencies as well as undertaken follow up groups.

3

Management in Education

Concept of Management

Management is concerned with both objective and subjective phenomena. Management involves values, attitudes, techniques and behvioural patterns at both strategic and tactical levels. All these, however, do not constitute an indivisible whole, which may be either accepted or rejected. They are varying in nature and are as much a part of the infinite variety of reality as anything else. Different types of organisation call for different types of management, different situation within a particular organisation require specific skills.

Management is the organ of the organisation. It has no existence in itself. Management divorced from the organisation it serves, is not management. There are three tasks which management has to perform, to enable the organisation to function effectively :

- Stating the specific purpose and mission of the institution. An institution exists for a specific purpose and mission. In business enterprise, this means economic performance.
- Making work productive and the worker achieving. An enterprise has only one true resource : human being. It performs by making human resources productive. It accomplishes its performance through work. Organising work according to its own logic is the first step. The second and far more difficult step is making work suitable for human beings. Management has to organise work for productivity, it has to lead the worker towards productivity and achievement.

management idea were able to grow and subsequently to flourish in both the United States and in Great Britain.

Summary of Historical Periods

Period	Representatives	Period of original development
Rational-Economic	Taylor, Fayol, Weber	1890-1930
Social	Mayo	1920-1950
Psychological	Maslow, McGregor,	1940-1950
	Bennis, Walton	1960-1970
Entrepreneurial	Peters, Kanter, lacocca	1980-1990

(*Source* : Huczynski, *Management Gurus*, London & New York, Routledge, 1993.)

Rational-economic period. The convictions of this period in the United States held that human's actions were aimed at maximising their self-interest. Human-kind was held to be primarily motivated by economic factors, to secure for them the greatest economic gain. The economic incentives were controlled by the organisation. The employees were passive agents, responding to economic laws, could be manipulated, motivated and controlled by the organisation. Their feelings were considered irrational and should be prevented from interfering with the rational calculation of self-interest. Organisations had to be designed in a way which neutralised people's irrational feelings and unpredictable traits. The closing decades of the 19th Century witnessed unprecedented changes in the social and economic realms of the United States. The creation of a high wage economy created a large potential market for mass-produced goods, which set the economic context for the early development of the large corporation. The existence of mass production provided the basic rationale for the application of systematic management systems such as Taylorism and later, Bedeauxism and Fordism. The significance of Taylorism, at least in U.S., came partly from the need to reduce and keep down labour costs in the context of a high wage economy. Taylor was confident that efficient management would drive politics out of the workplace. The spread of Taylorism coincided with the early and dramatic growth of concentrated large-scale production at a point when the

American Capitalist system exuded a high degree of confidence. As an ideology and philosophy of labour relations, Taylorism's utilitarian view of motivation was compatible with the American capitalism. The managerial ideas and institutions, which arose during this period, appeared to be applicable to society as a whole. Tylorism acquired an immense amount of ideological and political influence through a cult of efficiency which came to dominate practically every aspect of American life by the early 20th Century. Management techniques applicable to the firm were applied to the economy as a whole. Managerialism thus offered a set of organising principles, which came to shape every dominant U.S. Institution. The 'Science of Management', as expounded by Taylor and Henri Fayol (1841-1925), offered a set of ideas for controlling large organisations. These principles appeared to be as universal and immutable as the laws of physics. The 1880-1920 period saw speedy industrialisation in U.S., which was accompanied by a set of ideas emphasising individual enterprise, self-help and the opportunities offered by industrial and economic expansion. During this period management was forced to consider labour as a problem. Employers developed common principles of management policy and recognised the use of more refined techniques. By 1910, the primary emphasis had been placed on the 'man-management problem'. Methods were recommended for increasing labour productivity. Stress was placed on the idea that it was in workers' interests to collaborate with their managers' plans and to seek advancement in the enterprise by improving the quality and quantity of their output. The period itself was heavy with the legacy of 18th and 19th century social and natural philosophers, such as Adam Smith, Robert Malthus and Charles Darwin. All these writers, in their ways, had seen the world as operating upon certain principles. There had been the principles of the division of labour, of population expansion, of non-intervention in economic affairs and in natural selection.

Social period. Taylor's thrust of scientific management generated its own counter-ideology in the form of human relations ideas. Human relations ideas were based on the writings of Elton Mayo and the Hawthorne Studies. Human relations basically held that industrial society was a shaky fabric. Its scale, diversity and constant change means that it frustrated the basic human desire for intimacy, consistency and predictability in social loving. People looked for this in their jobs. Employees needed to focus on social relationship on the job. People were held to be more responsive to the social forces of their peer group than to the controls and incentives of management. Employees' responsiveness to management was seen to depend on the extent to which a supervisor could meet their social needs and needs for acceptance. Mayo's studies initiated a sociological orientation to work relations. Mayo and the human relations theorists introduced "Social Man", motivated by a desire to form relationships with others.

Psychological. The behavioural scientists, such as Argyris, Maslow and McGregor believed that "Self-actualising man" was a more accurate concept for explaining human motivation. According to Maslow, the needs people are motivated to satisfy, fall into a hierarchy, at the bottom of which are 'physical and safety needs' and at the top are 'ego needs' (the need for respect and recognition, for example) and 'self-actualising needs' (the need for personal growth). In general, lower level needs must be satisfied before higher-level needs can be met. Since many lower-level needs are routinely satisfied in contemporary society, most people are motivated more by the higher-level ego and self-actualising needs.

Behavioural scientists have made enormous contributions to our understanding of individual motivation, group behaviour, interpersonal relationships at work, and the importance of work to human beings. Their findings have helped managers to become much more sensitive and sophisticated in dealing effectively with subordinates. Behavioural scientists continue to offer new insights in important areas as leadership, conflict resolution, the acquisition and use of power, organisational change and communication.

Entrepreneurial. The entrepreneur, who sees environmental change as an opportunity, uses the factors of production to produce new goods and services. Entrepreneurship is different from management because it focuses on initiating change, although the entrepreneur must, at times, assume the functions of both manager and capitalist. Entrepreneurship can occur when an individual or group of individuals start a new business or it can occur within an existing corporation in a process called intrapreneuring. Entrepreneurship is an important area of study and contributes to a society's economic growth, its productivity, and to its supply of technologies, products and services. A nation's level of entrepreneurship depends on both economic factors (the supply and availability of capital) and non-economic factors (especially cultural and social differences that affect business).

The most important theory of entrepreneurship's psychological roots was put forward by McClelland, who observed that certain types of people had high need-achievement and that certain types of societies tended to create high need-achievements. The entrepreneurial process is sparked by some environmental change, which gives the entrepreneur an idea for a new product or service. Many ideas grew out of the entrepreneur's work experience; others came from the entrepreneur's observation of established businesses. In preparing to launch an entrepreneurial organisation, would-be entrepreneurs need to prepare a business plan. They must come up wit a business strategy, and choose a form of organisation. The entrepreneur must recognise the differences between the entrepreneurial and administrative cultures, choosing procedures, that will tap the best of both the cultures. In entrepreneurial companies, entrepreneurs are role models. Entrepreneurial companies always look for the people and units that do better and do differently. They single them out, feature them and constantly queries of their success. This open attitude helps keep entrepreneurial spirit alive.

Need for Management on Education

Educational institutions operate in an environment of constant change. At least a quarter of their student population change every year; faculty and staff do not stay the same forever. Since it is impossible to reach the pinnacle of service excellence once and for all, educational institutions must constantly identify and implement improvements in their set-up. Curricula need updating and revision, procedures need to be upgraded in accordance with new needs and technologies and the employees — faculty, staff and administrators need access to constant training and development opportunities. The process of continuous improvement requires a mind-set of constant monitoring, problem identification and research.

Educational institutions provide valued services to their students and aim to meet customer needs and expectations. While many institutions want to be responsive and effective, they are not sure of what responsiveness is or how to achieve it. Some institutions, however, seem unresponsive as they are concerned with other things than customer satisfaction. They are bureaucratic, rigid and reject innovation. Highly responsive institutions, on the other hand, survey current customer satisfaction and also look for the ways to enhance their services. Effective responsive institutions deliver customer satisfaction. They understand that customers seek value for their expenditure of time, effort and money and they try to increase value by operating efficiently and effectively. The institutions look at service quality, measuring customers' judgments of their service quality through complaint and suggestions systems and the feedback from these systems. Some techniques for appraising service quality include gap analysis, root cause analysis, performance analysis and benchmarking with other educational institutions, business and other organisations. With the results of the service-quality analysis, the institutions must take action to close the gap between what is currently being done and what the institution aims to become. The institution must determine which gaps are most important to customers and therefore most worth the time and effort to reduce. Areas for action include redesigning services, enhancing technical supports for service (*e.g.* with computers, training, tele-communications facilities etc.), training customers to help themselves and communicating with them. Service quality is not achieved once and for all, but must be preserved and enhanced through efforts at continuous improvement. Institutions that care about quality and produce quality, have identified the critical processes to generate a high level of customer satisfaction. They continually collect data to guide productive planning and action. According to Cubberley (1916), education had to have its corps of efficiency experts like 'every manufacturing establishment that turns out a standard product or series of products'.

Performance is the ultimate test of any institution. In an educational institution, where people want to serve a cause, you always have the challenge in getting people to perform so that they grow on their own terms. Performance in education occurs along three dimensions : (a) Knowledge, (b) as a participating citizen to perform within the economy, (c) able participation in the cultural life of society.

The school is organised on the assumption that the student is a thing to be worked on. The school has to be focused on performance and results rather than on rules and regulations and therefore needs a clear definition of its mission and it also needs a system to accomplish that. Basically, it is to keep an eye on the fundamental, long-term goal. One should start with the mission, which will guide through the endeavour. The moment one looses sight of the mission, resources are wasted. The main task of management is to ensure that all resources are used to the full. Obviously, technical rationality is a critical measure of management performance — reflected in costs, prices, profits, market shares and other economic yardsticks.

The purpose of organisational rationality is to achieve and maintain systems of organisation co-ordination. Co-ordination is useful to the health of any organisation. "The tasks of management are the reasons for its existence, the determinants of its work, and the grounds of its authority and legitimacy." (Peter F. Drucker, *Management* : *Tasks, Responsibilities, Practices*, New York, Harper & Row, 1974.)

Management is the organisational process of formulating objectives, acquiring and committing the resources and ensuring that the objectives are actually reached. It is a continuous process through which members of an organisation seek to co-ordinate their activities and utilise the resources in order to fulfil the various tasks of the organisation as efficient as possible.

APPROACHES TO MANAGEMENT IN EDUCATION

Educational Management is a field of study and practice concerned with the operation of educational institutions. There is no single generally accepted definition of the subject as its development has drawn heavily on several firmly established disciplines including Sociology, Political Science and Economics.

According to Glatter (1979), educational management studies are concerned with the internal operation of educational institutions, and also with their relationships with their environments, that is, the communities in which they are set, and with the governing bodies to which they are formally responsible. Glatter's statement delineates the boundaries of educational management but leaves open questions about the nature of the subject. Most writers on educational management emphasise the significance of purposes or goals in education. Culbertson (1983) claims that "defining purpose is a central function of administration". Cyert (1975) argues that "an organisation is developed to achieve certain goals or objectives by group activity". The centrality of the goal orientation of schools and colleges is common to most of the different theoretical approaches to the subject. There is disagreement, though, about the value of formal statements of purpose, about 'whose' purposes may become the objectives of the organisation and about 'how' the institution's goals are determined.

Certain approaches to educational management are concerned predominantly with organisational objectives while other models strongly emphasise individual aims. There is a range or opinion between these two views. Gray (1979) suggests that the management process is concerned with helping the members of an organisation to attain individual as well as organisational objectives within the changing environment of the organisation. It may be that individual and organisational objectives are incompatible or that organisational aims satisfy some individual aspirations. It can be assumed that most teachers want their school or college to pursue policies which are in harmony with their own interests and preferences. Coulson (1985) asserts that the goals and motivations of all the individuals involved in the life of the school, interpenetrate with the goals, officially attributed to the organisation.

The origin and development of educational management as a field of study have been well chronicled by Culbertson (1980), Bone (1982) and Hughes (1984). It began in the United States in the early part of the 19th Century. The growth of the subject reflects the view that the management of educational institutions is enhanced by training just as teachers require training to be effective in the classroom.

There are several major areas in which the management of educational institutions differ markedly from the management of other organisations.

1. The 'Objectives' of educational institutions are much more difficult to define than the purpose of commercial organisations. There are no clear-cut educational equivalents to such major private sector objectives as profit maximisation, output maximisation or product diversification. Schools and colleges are expected to develop the personal ability of individuals, to inculcate the accepted values and beliefs, to prepare pupils for the next stage of education, for employment. The teacher is expected to be the social worker, the custodian and the quasi-parent as well as developing the pupils' ability to learn. The ordering of priorities within this complex set of objectives is fraught with difficulties.
2. It is difficult to 'measure' whether or not objectives have been achieved in education. In commercial organisation it is possible to measure success in financial terms — sales have increased, profits are up, dividends are higher. In educational institutions any assessment has to be long-term to allow for the sheer length of the educational process. Still certain aspects cannot be adequately measured. How do we assess the extent of individual development? There is a danger that educational institutions will be assessed only in terms while the more elusive criteria are undervalued or overlooked. Handy (1984) suggests that the absence of acceptable bases for evaluation creates serious management difficulties in educational set-up. Without clear and agreed objectives there are no criteria for deciding how to allocate resources. Without clear measures of success there are not obvious ways to assess the progress of individuals. Every judgment becomes subjective.
3. In educational scenario pupils may be regarded as clients or outputs of schools and colleges. As

participants in a production process these young people differ markedly from the raw material of industry and commerce. Children cannot be processed, programmed or manipulated. The learning process is built on personal relationships with all the idiosyncrasy and unpredictability that implies. This human variability reinforces the problems of measurement, as discussed earlier.

4. The managers and teachers in schools and colleges are from a common professional background with shared values, training and experience. As professionals, teachers claim a measure of autonomy in the teaching and learning process. It suggests that teachers should be able to participate in school or college decision-making because their commitment to the implementation of decisions is essential if the decision-making process is to be more than a ritual. The 'client' relationship between teacher and student differs in several respects from other professional-client links. Teachers have regular and extended contact with pupils; often several encounters — a week for a period of years. Moreover, pupils have little opportunity to select their teachers. Children are required to spend a good number of years as members of educational institutions and have to accept that their teachers will be chosen for them. This gives organisational problems in schools.
5. There is a fragmented organisational and management structure both within and impinging upon educational institutions. The climate for school and college decision-making is strongly influenced by a plethora of external agencies and group — politicians, officials and inspectors at state and local levels as well as parents and both formal and *ad hoc* pressure groups. There are multiple decision points inside the institution concerning the management of the school or college and its sub-units.

The organisational structure of a school is established in order to discharge the expectations and duties imposed on it by the society it serves, as well as to realise the goals of the teachers involved. The appropriate operational and functional roles and relationships in a school are usually formulated rather than formalised in such a way that the abilities and inclinations of staff are used to best advantage. A school is an organisation primarily concerned with the interactions between its members — teachers, pupils, office staff etc. and these interactions produce a social climate peculiar to each school. The school climate is the name given to the atmosphere which develops from the ways in which the head teacher and staff work among themselves and with the pupils, which can be assessed in several ways depending on the priority given to the factors which influence it. These factors include (a) the spirit of co-operation and teamwork with which tasks are undertaken; (b) the effectiveness of the communication system; (c) the degree of commitment and loyalty of its members — teachers and pupils; (d) the effectiveness of conflict resolution; (e) the extent to which participation in decision-making takes place; (f) the degree of mutual confidence between head and staff and between staff and pupils.

Halpin (1966) has stressed the importance of the head teacher's role and suggests six school profiles which can develop as a result of social and organisational behaviour. In an 'open' climate, the head teacher is constructive and flexible. There pervades a feeling of achievement in the working atmosphere and good personal relationships develop. Where the head teacher remains aloof, impersonal, arbitrary and firm, leadership is lacking, the climate is described as 'closed' and teachers gain little job-satisfaction or personal fulfilment from their work. In an 'autonomous' climate, teachers are given a fairly free hand in their work, which increases the sense of job-satisfaction, but social needs receive little consideration. An authoritarian head teacher may exert considerable control over the staff and creates a more 'controlled' atmosphere, which is appreciated by many teachers. A familiar climate, however, is produced when the head teacher appears more concerned about social relationship than in giving firm leadership resulting low achievement, with the result that morale suffers. The 'Seductive over-solicitousness', as Halpin puts it, of the head in a formal climate is distrusted and there is little incentive for effective task completion among staff.

(a) Early Approaches to Management

While studying Case Approach, we are reminded of Henry Ford. In large measure, the managerial approach of Henry Ford is a paradigm of much that was constructive and much that was imperfect in early approaches to management. Ford's management style was the darker side of his innovative personality and it led to chaos in the ranks of management. Ford did not take criticism well, specially of the Model T. When his top engineers suggested changes, they were ignored and some were even fired. As a result, his rivals were able to make a great deal of headway in the automobile market. Ford also liked to play his subordinates against each other, to the point of assigning two of them the same title and seeing which one would back down first. Competent managers left the firm, which remained overly dependent on Ford himself. However, Henry Ford left an indelible mark on both American industry and society. His name is synonymous with mass production and the development of modern management theory. Ford is also important as a management theorist, who developed ideas about how organisations function.

The Interpersonal Approach reminds us of Mary Parker Follett (1868-1933), who was convinced that no one could become a whole person except as a member of the group. Thus she took for granted that labour and management shared a common purpose as members of the same organisation, but she believed that the artificial distinction between managers and subordinates obscured this natural partnership. In her behavioural model of organisational control, control was sponsored by and oriented toward the group. Self-control was exercised by both individuals and groups, with the result being shared of control of power. Follett characterized the interactive or integrative nature of self-control groups as a "holistic" model of control.

In the Group Behaviour Approach, a group is defined as two or more individuals, interacting and interdependent, who come together to achieve particular objectives. Group can be either formal or informal. By 'formal', we mean a group that is defined by the organisation's structure, with designated work assignments, which are directed towards organisational goals. As for 'informal' groups, they form naturally as responses to the need for social contact. Groups can be sub-classified into command, task, interest or friendship categories. The Command Group is determined by the organisational chart. It is composed of the employees who report directly to a given manager. An elementary school principal and her twelve teachers form a command group. Task Groups, also organisationally determined, represent persons working together to complete a job. However, a task group's boundaries are not limited to its immediate hierarchical superior. For example, if a college student is accused of a campus crime, the investigation may require communication and co-ordination among the dean of academic affairs, the registrar and the principal of the college. Such a formation would constitute a task group. It should be noted that all command groups are also task groups but because task groups can cut across the organisation, the reverse need not be true. People who may or may not be aligned into common command or task groups may affiliate to attain a specific objective with which each is concerned. This is called interest Group. Employees who band-together to have their vacation schedule altered, to support a peer who has been fired or to seek increase fringe benefits, form a united body to further their common interests. Groups often develop because the individual members have one or more common characteristics, which are known as friendship groups. Informal Groups provide a very important function by satisfying their members' social needs. These interactions, though informal, deeply affect the individuals' behaviour and performances.

Groups differ in their cohesiveness — that is, the degree to which members are attracted to each other and are motivated to stay in the group. Some work-groups are cohesive because the members have spent a great deal of time together; sometimes the group's small size facilitates higher interaction. Cohesiveness of the group is important because it has been found to be related to the group's productivity. Studies consistently show that the relationship of cohesiveness and productivity depends on the performance-related norms established by the group. The more cohesive is the group, the more its members will follow its goals. If performance-related norms are higher (*e.g.* high output quality work, co-operation with individuals outside the group) a cohesive group will be more productive than a less cohesive group. But if cohesiveness is high and performance norms are low, productivity will be low. If cohesiveness is low and performance norms are high, productivity increases but less than in the 'high cohesiveness-high norms' situation. Where cohesiveness and performance-related norms are both low, there will be no significant effect on productivity. In order to encourage group cohesiveness one might (a) make the group smaller, (b) encourage agreement with group goals, (c) increase the time members spend together, (d) increase the status of the group and the perceived difficulty of attaining membership in the group, (e) stimulate competition with other groups, (f) give reward and recognition to the group rather than to members or (g) physically isolate the group. A common stereotype about groups is that the sense of team spirit spurs individual effort and enhances the group's overall productivity. Research findings show that more people may be better in the sense that the total productivity of a group is greater than that of one or two people, but the individual productivity of each group member declines. The conflict arising from heterogeneity of the group can be a particular problem when management creates teams of employees for varied functions in the organisation.

(b) The Systems Approach

The Systems Approach views the organisation as a unified, purposeful system composed of inter-related parts. This approach gives managers a way of looking at the organisation as a whole and as a part of the larger, external environment. The systems theory tells us that the activity of any segment of an organisation affects, in varying degrees, the activity of every other segment. The parts that make up the whole of a system are called 'subsystems'. And each system, in turn, may be a subsystem of a still larger whole. Thus a department is a subsystem of a college or an institution, which again may be a subsystem of a university. 'Synergy' means that the whole is greater than the sum of its parts. In organisational terms, 'Synergy' means that as separate departments within an organisation, co-operate and interact, they become more productive than if each had acted in isolation. For example, in a small firm, it is more efficient for each department to deal with one financing department that for each department to have a separate financing department of its own.

A system is considered an 'open system' if it interacts with its environment. It is considered a 'closed system' if it doesn't interact with its environment. But the extent to which they do so, varies. For example, an automobile plant is a far more open system than a monastery or a prison. Each system has a boundary that separates it from its environment. In a 'closed system', the 'system boundary' is rigid; in an open system, the boundary is more flexible. The system boundaries of many organisations have become increasingly flexible in recent years. For example, oil companies wishing to engage in offshore drilling now must have concern for the environment.

A system has 'flows' of information. These enter the system from the environment as 'inputs', transformation processes within the system (operations that alter them), and exit the system as outputs (goods and services). 'Feedback' is the key to system controls. As operation of the system proceeds, information is fedback to the appropriate people or perhaps to a computer so that the work can be assessed and, if necessary, corrected.

Systems theory calls attention to the dynamic and interrelated nature of organisations and the management task.

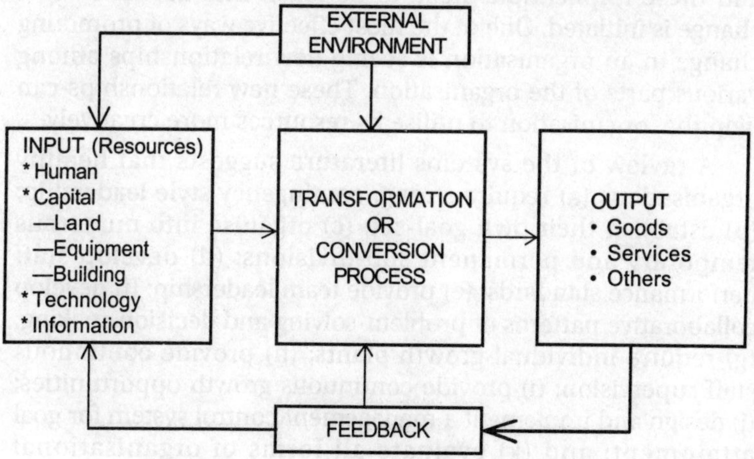

The Flows and Feedback in an Open System

Thus it provides a framework within which we can plan actions and anticipate both immediate and far-reaching consequences, while allowing us to understand unanticipated consequences as they develop. With a system perspective, managers can maintain a balance easily between the needs of the various parts of the enterprise and the needs and goals of the whole organisation.

Systems theory suggests that an important task of an administration is to provide leadership in the integration of the system, both internally and with external and super-ordinate systems. Such integration implies the need for differentiating the function of each system and subsystem, and then providing for the development of relationships between and among them. The traditional approach has been to provide for specailisation on the part of the individual and the organisational unit, with the result that in most instances the organisation has tended to become fractionated. The advantages of thinking in terms of systems and subsystems is that administrators can focus attention upon the inter-relatedness of the objectives, activities, and resources, represented by the various individuals and units in the organisations.

Because of the interrelationships in a system, change in a school organisation needs to be viewed in terms of the organisation as a whole. Change in any one part of an organisation, has implications for the rest of the organisation, and these implications need to be taken into account when change is initiated. One of the most effective ways of promoting change in an organisation is seeing new relationships among various parts of the organisation. These new relationships can help the organisation to utilise its resources more creatively.

A review of the systems literature suggests that healthy organisations (a) require strong, contingency style leadership; (b) establish their own goal-set; (c) organise into numerous temporary and permanent sub-divisions; (d) develop staff performance standards; (e) provide team leadership; (f) develop collaborative patterns of problem-solving and decision-making; (g) require individual growth plants; (h) provide continuous staff supervision; (i) provide continuous growth opportunities; (j) design and implement a management control system for goal attainment; and (k) evaluate all forms of organisational performance. Systems concepts about organisations are descriptive of the ways in which functions, tasks and roles — all related to each other. A life-flow exists in healthy systems in which all forces are working towards a common goal and are constantly responding to feedback, needs and aspirations.

Education as a System

Developed communities are characterised by the existence of numerous social structures of which the family, state, school associations, clubs are the important ones. Each of them has its own particular structure, nature and existence because they were brought into existence to carry out particular functions in the community. Man lives in and through the social structures, that he established to meet his particular needs. For example, the family comes into being because of man's desire and ability to procreate. The family is therefore a social structure, biologically founded and ethically qualified; and parents and children are essential to its composition. All societal structures are based on one or more human functions. This is so because it is only man that creates culture and therefore deploys only those social structures which he needs. His needs and sense of responsibility, in turn, are determined by the level of development and the nature of his functions, for example, if he doesn't have a need for education then schools, colleges and universities are not established. The education system is composed of a great variety of social structures. The education system is different from the more usual social structures; in that it is a combination of social structures that are concerned with education. Stone (1974) regards it as an interrelated structure, that is, a social structure that consists of a great variety of social structures that are interrelated with one another so as to function as an integrated unit.

The education system is much more than the school. In a community's education system, the family, school, school committee, education department and the state — each have a particular task. The education system is not the sum of the above-mentioned structures but is an intertwinement of them. It forms a unit with various components. The education system appears out of the totality of many social structures as a unified structure with its own particular identity and nature. It is an interrelated structure in contrast with the family, school committee and even the school, which are all single-type structures. Within the education system there are various social structures such as the state, families, teachers-in-organisation, commerce and industry and many others that are interwoven with educational institutions for the purpose of which they were established.

The education system is directly concerned with that which is pedagogical. According to Stone (1974), the education system is a social structure that is qualified by educative teaching, is typified by and finds its purpose in it. Although there are a large number of non-pedagogic social structures involved in the intertwinement, the education system, in all its activities, is qualified by pedagogic principles. As Klafki (1973) observes, as a pedagogically qualified social structure, the education system is concerned with the creation of a previously formed educational field for the accelerated, planned and normative reservation of the child to his world and the world to the child with a view to his cultural mandate in the world. Because it basically concerns the child, the education system finds its limits and possibilities in the nature of the child.

Sergiovanni and Carver (1980) describe the educational system as composed of three connected sub-systems : political, organisational and human. The political sub-system makes decisions about systems-inputs to be converted to outputs through policies and legislation. The organisational system is composed of a formal unit with its own survival goals of eliminating uncertainty and maintaining the present equilibrium and any additional goals and purposes that society gives it. The organisational system works through people (the human system) to achieve its goals and purposes. It is hierarchical in structure, and each of the three major sub-systems includes its own structure.

PLANNING IN EDUCATION

Educational planning is a relatively new concept being utilised in more sophisticated school systems. Originally introduced as a result of planning experiences and models in the military in U.S., the planning role is rapidly being developed to resolve educational problems with a set of tools, unique to education. Thus the Planning, Programming, Budgeting System has been refined to the Planning, Programming, Budgeting, Evaluation System or PPBES.

The development of an educational model to carry on the planning function must rest on the validity of several assumptions : (a) The planning role must be separated from the operational role so that resources are available to carry out the planning task. Those charged with planning obligations cannot devote the time and energy needed for the operational duties of the system; (b) Planning encompasses the total system and its needs; (c) The diversity of skills needed amount a planning function mandates a "team" effort; (d) Resources needed for planning are crucial to its success and are a continuing part of any budget.

Among numerous planning traditions, five classifications that can complement organisational processes are : synoptic planning, incremental planning, transactive planning, advocacy planning and radical planning; synoptic planning is the centre-piece or locus of planning activities.

Synoptic planning involves centralised control of decision-making more than the other four classifications. This approach may be best characterised as feasible for translating policy implications and adapting to problems and social settings. With its major emphasis on producing plans, the synoptic approach is goal-oriented, relying heavily on deterministic and probabilistic models with some acknowledgment of the judgmental process.

APPLICATION OF MANAGEMENT CONCEPTS IN ACADEMIC AREAS OF THE EDUCATION SYSTEM

Curriculum Development

The school is an organic community and the curriculum is the comprehensive range of student learnings for which the school takes responsibility in the pursuit of educational goals and values. Accordingly, school improvement must focus on the whole curriculum and the whole child. In performing its role as an active agent of curriculum construction, the school must seek new ways of understanding and relating to profound changes in national educational policies and structures in the culture of modern society. To respond creatively to these challenges and play its part in these movements is increasingly being recognised as a necessary condition for the growth and reform of schooling.

Curriculum is defined as the plan of learning for achieving the intended outcomes of education. Curriculum development is the process of designing and planning the curriculum. The curriculum includes both formal and informal learning activities in relation to planned courses of instruction. The curriculum functions through learning experiences and instructional materials from various fields of knowledge. Learning experiences are provided that meet not only the general needs but also the special needs related to the unique abilities, interests and expectations of each individual student. The curriculum needs to reflect a pluralistic viewpoint. A pluralistic curriculum is defined as one in which the diversity of cultures, including history, languages, lifestyles and patterns of beliefs, are interwoven throughout the programme. Instructional activities are planned and implemented to stimulate growth in knowledge, understanding, attitudes, ideals, habits and skills, appropriate to a full life based on the democratic and moral values of the society.

Professional Growth — In-service Training

Promoting continuing professional development of staff must be a priority for any organisation which aims to raise performance standards. Improvement in staff performance can be thought of virtually synonymous with in-service education or training. The performance of individual staff members tends to be most responsive to change and improvement through in-service education. In-service education or in-service training is used to mean any planned programme of learning opportunities of staff members of schools, colleges or other educational agencies for purposes of improving the performance of the individual in already assigned positions. The purposes of in-service training or education are clearly restricted to learning outcomes related to the improvement of performances of the staff, that is, of various staff groups in the organisation. The term 'staff development' is concerned with staff and is therefore concerned with both the needs of individuals and those of the organisation as a whole. It opens up the possibility of a much wider range of activities.

A planned programme is specified, contributing to its purpose. In prescribing in-service education as planned, the emphasis is placed on designing learning experiences, assessing needs, projecting expectations, budgeting, assigning responsibilities and evaluating. Learning opportunities suggest the uniquely educational character of in-service education, distinguishing it from curriculum development, staffing, public relations and other tasks that may well be educative but have quite different goals and objectives. Staff members are specified as the persons for whom in-service educational activities are planned. In school settings, most in-service education should be targeted to teacher improvement because teachers are the bulk of the personnel employed. In agencies other than schools, in-service education is properly targeted to any staff group that is crucial to the effective operation of the agency. The term 'performance' is utilised to emphasise job relatedness and practice. A narrow interpretation of this term would emphasise training activities and skill development. The 'individual' is designated as the focus of in-service education. Only individuals can learn. However, the specification of individual performance as the focus :

— to treat the staff as professional people relying fundamentally on their own standards, and devotion to duty.
— to exploit the strengths, recognising that everyone has strengths and weaknesses.
— to respect the diversity of approach and personal idiosyncrasy, recognising that these add to the richness of living.

In-service Training

Learning by experience, always implies experiences that are real or at least clearly associated with a concrete reality, in contrast to an abstraction or fragmentation, remotely related to reality. Butts (1967) developed and tested in extensive fashion a "classroom experience model" of in-service training in implementing teaching strategies. The micro-teaching programmes initiated for pre-service training at Stanford University in U.S. were later extensively utilised for in-service education (Meier 1968).

On-the-job training is an old and respected concept and has long been a standard practice in some non-academic fields. There appears to be growing interest in employee development in many fields of business and industry that emphasises on-the-job methods. Professionals have also made extensive use of reality as experience for in-service training. The medical residency is perhaps the classic example of intensive field experience utilised for training beyond the terminal degree. There have been numerous recent efforts to provide new skills for staff by direct guided experiences in a school setting.

Because learning is central to educational change and because professional development comprises the typical format through which teachers are expected to learn, professional development becomes a pivotal component in school change. Yet typical professional development offerings fall short of these responsibilities. Teachers do not view their professional

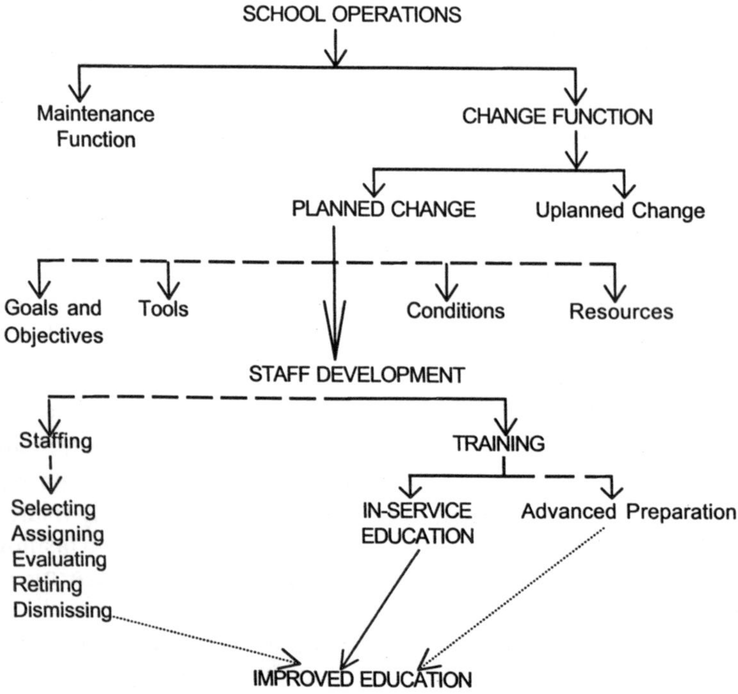

In-service education as a part of staff development

development opportunities as particularly useful. The analyses of the conditions of professional development of teachers suggest that the complexity in teachers' working conditions (*e.g.* conflicting goals, teacher-student relationships, the nature and consequences of teacher interactions etc.) requires a greater ranger of support than the intellectual content found in typical in-service training. Moreover, most staff development opportunities are hindered by workload, time and teaching responsibilities and are typically content or pedagogy-free, and thus offer small benefits to teachers. Smith & O'Day (1991) discussed teacher in-service opportunities as severely limited in scope and duration. They are rarely of sufficient depth and scope to give teachers the experience necessary to make major changes in their approach to instruction.

Evaluations should attempt to determine the relationship between what is done in the programme processes and the changes observed in teachers' outcomes. The two most important evaluation goals are :

(a) to determine the extent to which the programme was implemented and

(b) to identify the factors that can maximise programme implementation in the future.

Thus, process and outcome evaluation questions should be analysed carefully in search of the patterns and relationships between programme activities and observed changes. It is also suggested that discrimination be made between 'impact' and 'improvement', when looking at outcomes. 'Improvement' involves a judgment about whether or not something is better. 'Impact' involves the more limited questions of whether or not something is different. An observed difference may or may not constitute improvement depending on who is making the value judgment about whether or not a change is for better or worse. The empirical question here is not improvement, but change. One is to begin with the question of whether or not they are different. Has the programme been effective in changing teachers? Can they do things now that they could not do before? These are empirical evaluation questions. Data from these evaluation questions can be used to determine whether or not such changes and differences constitute progress or improvement. The evaluation questions that focus on how things are different reflect an 'impact' evaluation purpose. The purpose is to find out the programme's actual effects on teachers. Did it effect teachers in terms of knowledge, skills and behaviour? Evaluation questions that focus on how things are better reflect an 'improvement' evaluation purpose. The purpose is to find out, 'does the observed impact constitute improvement'?

In-service training should be a systematically planned investment in the development of knowledge, skills and attitudes needed by an individual to perform a job to a satisfactory level. Ideally, training should be followed by putting into practice what has been learnt. It is always important to ensure that the arrangements for the prompt application of knowledge and skills are included in the training plan.

Every organisation needs to measure and evaluate the performance of its people. Performance measurements and appraisal systems include determining what performance to measure, the rating or evaluation of performance, the on-going management of performance, reward allocation and development.

Scope of Performance Measurement, Management & Appraisal

Procedures for Evaluating the Staff Development Programme

The programme of staff development needs to be evaluated in terms of criteria of its worth, grounded in its own purposes. A major purpose of the programme is to establish a schema of supervision that will accentuate the professional development of staff members. Another major purpose is to highlight staff development so that it becomes a predominant function. Still another is to create, by means of the programme, significant improvements in the knowledge, understandings, skills, and attitudes of the staff for whom the programme is designed.

For evaluating the programme, an evaluation design is to be created. It can be experimental, so that changes in individual staff members who participate in the programme can be gauged with reference to changes in a controlled group of staff members with whom they are at least roughly paired. The goals and objectives should be very clearly stated. Accordingly, evaluation means are to be chosen to evaluate the attainments. The instruments chosen (tests, inventories, observation devices and instruments for artifact analysis) may be borrowed from other sources to fit criteria, goals and/or objectives. These should be appropriate, valid, reliable and capable in interpretation. The programme should be evaluated formatively (while it is in the process of development) and also summatively (at or near its completion). The results of formative evaluation may cause its modification while it is in process.

4

THE STRATEGY

- For the stiff fiber to be used for cordage and rope through Financial Corporation.
- For reconstruction of small farms, solar energy systems, food, fruits, fodder plants, trees, the trained rural leaders.
- For farm families towards sound nutrition, scientific agriculture and animal husbandry with the 'Host of community partners'.
- For the establishment of agriculture family, school offering for basic education grounded in rural realities with new resources.
- Today training for tomorrow, and hope for future of its youth for beginning to bloom.
- To fill position in Information and creative services.
- Tree scientists/horticulturists.
- Regional coordinators.
- Ecologists for the National Resource strategy and policy.
- Leader of the ecosystem Rehabilitation programmes.
- To prepare female entrepreneurs to face in the world of global changes.
- Changing world market; altering social structure, accelerating development in information technology.
- Expanding industrialisation.
- Extending social services and education.

Mission for the Tasks

Enhance effective participation of women in industrial development through promoting of women's ventures that efficiently utilise their potentials, skills and capabilities.

Tasks

- Provide technical and managerial assistance to female entrepreneurs through advisory services, guidance and industrial information.
- Training.
- Create effective coordinated programmes with other women.
- Support entities.
- Conduct studies and research.
- Promote self-employment programmes.

International centre for research will function :

- To improve human welfare by reducing poverty, increasing cash income.
- Improving food and nutrition security.
- Enhancing ecosystem resilience in the tropics through improved agro-forestry systems.

Values

- Dynamic ecological and global prospectives of people, contexts and events.
- Knowledge that improves daily lives of current and future generations.
- Creativity and critical thinking.
- Access to information, education and technology.
- Collaborations.
- Excellence, integrity and professionalism.
- Building on strengths of students and programmes.
- Diversity of people and ideas.

PRIORITISATION OF RESEARCH AGENDA

The new research initiative and collaborative research for conducting training programmes should be developed, in view of — world-wide wisdom, social responsibility, establishment of strong and influential networks of individual organisations that are engaged in research and grass-root activities to promote gender equity in science and technology.

Objectives

- To encourage research into all aspects of gender

differentiation in science and technology education and employment.

- To facilitate the entry of women into employment in the field of science and technology and their progress within such employment.
- To provide a forum for dissemination and discussion of research finding and experiences of those in the field.
- To provide a support network for those working towards objectives outlined above.

Issues

- Which scientific and technological advances would be liked to be seen in future?
- Which applications facilitate by the current state of science and technology are desirable?
- What demand will be the answers to the above questions and make upon the formal education sectors?
- How can science and technology education be constructed to embrace diversity of gender, ethnicity, class, religion, etc.?
- How can informal and indigenous science and technology be acknowledged and integrated with formal science and technology?

The emerging demand of what are to be met through increased productivity, sustainability, environmental and ecological security, profitability and competitiveness in the backdrop of decline in factor productivity, loss of biodiversity, natural resource shrinkage and degradation, climatic change, intensifying competition in international trade and widening.

SIGNIFICANCE

For creation of vibrant new partnership — to reach the most disadvantaged and under-served and the most vulnerable — the young child. The emerging strategy focuses on innovation, fostering decentralised flexible responses to the care of young children, girls and women and reaching through integrated family and community based interventions for survival, growth and development of socially backward villages and urban slums. With the special preference to those areas which are predominantly inhabited by vulnerable and weaker sections of

Proposed Schedule of Courses for One Year Postgraduate Diploma in Human Ecology

Module I	*Module II*	*Module III*	*Module IV*
Agriculture and Environment	Extension education and Management	Farming system—Practices	Legal Environment
Nutrition and Health Related issues	Rural Sociology	Production and operation Management	Agribusiness, Post-harvest and processing, Biotechnology
Women and Development	Communication and Audio Visual for Development	Financial Management Managerial Economics	Biodiversity and environment Non-conventional energy systems
Computer basic	Management and Extension	Management of Formal and non-Formal enterprise	Optionals on Farm, off Farm, Non-farm sections
			Creche Religion

Duration of the course — One year

Working hours — 5 days a week with 8 hours per day

Total Maximum marks 1000

Module I — (4 written papers — 200 + Practical — 50)

Module II — —do—

Module III — —do—

Module IV — (Written paper — 50 + Project work — 200 marks)

For each written paper of 50 marks — 45 hrs. of teaching.

For each practical paper of 50 marks — 45 hrs. of laboratory work.

the society *i.e.* Scheduled Castes, Scheduled Tribes and low income families found in economically backward areas, drought prone areas and areas in which development of social services requires strengthening. It also believes that these goals are an integral parts of all efforts to achieve sustained and sustainable social and economic development that meets needs, ensures well-being and protects the natural resources on which all life depends.

The philosophy of human ecology provides the context for developing and serving the individuals and their families from a human environment perspective towards building and sustaining a productive quality of life. The scientific expertise, infrastructure facilities and research needs to be strengthened by bonding the relationship between research and educational institutions and industry in a symbiotic mode of conversion of research findings into products, processes and technologies for human resource development.

Interdisciplinary knowledge and skills to deal with the problems through vigorous research, active extension education programme will outreach the community. This will help them to meet the challenges imposed by the new economic and social order and improve their quality of life. The concept and view points developed here will be helpful in framing educational strategies in higher education and research needs in the field of human ecology.

The Second Congress on Human Ecology was held on 4-6th February, 2001 to foster discussion on present global policy issues and thrust areas through sessions featuring national and international subject experts. It will give an opportunity to the participants to interact with experts of the field to seek solutions on a wide range of themes :

1. Linking higher education with community service.
2. Changes required in the area of Home Economics and Home Science Programmes to make them relevant to all sections of the society.
3. Setting priorities in research to help community in the areas of Agribusiness, Biodiversity & Environment, Biotechnology, Non-Conventional Energy System and Human Resource Development.

4. Action Mechanism for Research and Development.
5. Partnership for sustainable development at all levels (local, nation, regional and global).

RESEARCH PRIORITIES

Since extension as a discipline has its objects, the rural and agriculture as an integrated whole, it has borrowed from various disciplines and certain assumption, theories and principles about the various aspects of human use with the special vocabularies of these disciplines. Thus sociology and rural sociology may emphasise, in studying human behaviour, their law of development of society. The anthropologist may stress the importance of culture, a psychologist may maintain behaviour aspics the psychologists and social psychologist may be deeply interested in group psychotherapy and providing the effectiveness of interaction upon the individuals through, feeling conditions and habits.

An economist may believe by nature that in extension education the dominant determinants and economic resources and technological skills. Apart from this there exists a number of disciplines such as political sciences, ethics, home economics and other social science which rightly claim that their point of view need to be inducted into the theoretical orientation so as to enrich the discipline. A model home science curriculum would therefore care to cover all such factor fields, interests which effect a person's development his life, in the family and in the community including the world of work.

There are series of tools and mechanisms which management can use to promote links among researchers and between researchers and technology transfer workers. Committees, liaison persons and coordinating units often have a difficult time creating effective links, unless they have power and authority or control a budget. We have to be very precise about what kind of links are needed, at what level they should be established, and what goals we hope to achieve. When we need so many linkage mechanisms that an organisation becomes excessively complex, we have to start thinking about reorganisation rather than creating more position or units in the organisation.

From its inception human ecology has focused human being relationship to the material goods and social and economic system necessary for living. Over the courses of its history, human ecology will seek and applied knowledge from many sources and disciplines helpful to individual and family in meeting everyday needs and addressing significant human environment problems. The focus has been and continues to be on the activities of everyday life associated with meeting recurring sustenance needs for food, clothing and shelter and human development and relationships within various environments such as the family work place market systems and other community settings.

Historically, a major focus has been on the micro environment — home and society. As society technology, and home science institutions have evolved, increasing attention has been given to other settings in which needs are met and to interdependence within and across family, home, work, business and institutional environment. The history provides a sense of the holistic and ecological nature of the home science colleges and shows the ways in which inter-related functions and components are connected.

Concern for the well-beings as and families. The field is defined as the study of human beings as an integrated whole, his man environment and the interaction between them. The focus has been the interaction of humans with their environment, emphasising in-depth study of the reciprocal effects that take place physically, socially, and aesthetically. Home economics had originally begun as a unified field for the application of science to everyday life in the home, family and community but over the years had developed into specialised fields and professions. This development reflected scientific, technological, socio-economic and political changes as well as changes in the family. Including greater attention to the development, selection and use of consumer goods by families and individuals.

The questions of methodology proper are linked both with the theoretical issues related to philosophy and to applied issues associated with various mathematical and quantitative techniques. At an abstract level, questions of method and questions of theory overlap and are often identical. They are related to various theoretical positions like positivism, phenomenology, structuralism, post structuralism and so on. To include various methodological orientations of the studies on Indian society and also to point out gaps and theoretical limitations of the specific class of methodological issues, the discussions on methodology posing of many of the questions inevitably leads to the problems of the very foundations of human knowledge. Thus in many cases the discussion is carried to the point at which the most recent philosophical issues are examined.

Volunteers, volunteer association and NGOs. The use of volunteers is an efficient way to work towards community objectives. Not only can government choose the activities that it wants to organise; it can also help foster a healthy sense of community. Government need not get directly involved. It may instead promote Volunteer Association or funds, NGOs or private firms that can organise activities, small scale organisations are a common example.

Expenditure and contracting. Expenditure consists of any use of public money such as contracting monitoring, investment, procurement, enterprise and public private partnerships, Government and public private partnerships. Government can work directly towards specific community objectives, or by offering contracts that serve those purpose. Furthermore traditional engineering approach often encourages expansion of infrastructure and supply augmentation rather than alternatives such as demand management. Still, investment in improved technologies, such as design of sewers, design and construction of main holes, inverted siphons protection of water supply, waste water facilities for sewage keeping with current usage for more conventional municipal waste water collection and treatment systems. Innovative approaches to collection and treatment particularly for the very small municipal systems practices for safeguarding of the public health or protection of water quality, for desirable procedure or methods with deviations subject to individual consideration. The wastewater treatment works for safety, design, essential facilities and laboratory. Settling sludge and scum removal, protective and service facilities, Sludge processing, storage and disposal,

Biological treatment systems, Disinfection — chlorine disinfection, dechlorination, ozone. Supplement treatment — handling and treatment of septage at a waste water treatment plant.

Innovative construction of a municipal garbage composting facilities, with a small-scale energy efficient sewage treatment plants and fish to change sewage into clear water. Any technology that encourages the recycling of resources should be seriously considered by communities that attempt to plan sustainability. The expenditure may be taken as solution to many policy problems.

Public private partnership. Joint development between government and institutions can enter partnerships with particular firms that either have resources that are valuable to the government or have enough employees to do useful community labour.

RESEARCH, EDUCATION AND OUTREACH AND LINKAGES WITH COMMUNITY SERVICE

The educational institutions have clear goals that are compatible with national policies and defined channels which users can make their demands known. It is still important to recognise that the tasks of producing and transferring relevant technology are carried out by two groups of people — researchers and technology transfer workers — and that these groups differ in attitude, interests and training.

The goal of human ecology is to provide interdisciplinary study focused on individuals, families, and their interaction with their surroundings both material and social with the identification, selection and use of resources by individuals, and families in order to provide the means for achieving optimum human development.

- Two years postgraduate programme in Human Ecology to prepare young men and women in managerial carriers in developmental organisation.
- M.Sc. Biotechnology, Microbiology, Forestry, Human Ecology, Real Estate Education.
- B.Sc. Medical Biotechnology, Medical Microbiology, Agriculture Forestry, Home Science, Human Ecology,

Microbiology, Biotechnology Physiotherapy, Mass Communication and Tourism etc.

- Pre-Service courses to equip all the graduate students for fulfilling necessary pre-requisites of rural and underserved population through integrated population research and use of community focus health education and public service programmes for 1 year duration.
- In-service courses and training programmes on agriculture education and life sciences leading to teacher certification will be developed in all collaboration with rural and undeserved population through integrated and education research for training in farm and off farm sectors for HRD.

It provides the forum of interaction between beings and their environments. It is a means to improve people's lives and environment with the focus on interactions of humans. This needs a framework for types of research is needed for understanding the basic principles of research methodology to discover relationships between existing non-manipulated variables. There is no single scientific method for scientists to carry out research methods and techniques to behaviour and instruments we use in performing research operations such as making observations recording data, techniques of processing data, like giving things in gathering materials, indexing them, participation in the field work required, and also training in techniques for collection of data, way of systematically solving the research problems. Research designs are needed for conceptual structure within, which research is conducted to analyse the situation to find out the change required in it. The approaches used are individual approaches, group approaches, and mass approaches. The quantity required of home scientist extension workers have to self-confidence, good judgment and also must have professional qualities.

Financial incentives. Financial incentives are an alternative to traditional regulatory instruments. They do not generally require enforcement because they create a constrained market environment in which firms normally as they normally would.

Evolution of research system. Expanding the research for pilot scale testing of the technologies already developed for commercialisation would be valuable to organisations within outside UGC system. This fund support would also help to broaden UGC base by drawing upon the strength of individual research institutions, which in term can fill gaps in unforeseen but critical areas of research related to regional problems besides contributing to developing staff skills through extension outreach and study tour — entry of institutions outside the ICAR/SAUS systems for enhanced inter-change with national and international centres of scientific innovations.

Several changes have taken place in the organisation and functional set up of Institute of Home Economics and in their research and development programmes. Commensurate in these, there has been no proper emphasis towards development of management skills of faculty members to make national research programmes more efficient and effective. Projects for R & D activities are time bound, and lay high priority to meet challenges such as poverty alleviation, food and nutrition security, natural resources conservation and overall sustainable development of agriculture and human resource; efficient managers are required.

Selection of species. Energy farms are still in the conceptual stage. It require detailed studies in genetic, agronomic and other aspects particularly the selection of appropriate sites and species, spacing and fertilizer usage and resistance to pest and diseases to maximise dry matter production and increase availability of alternative source of energy. Following criteria need to be adopted for selection of species, case of establishment and regeneration after harvesting, freedom from pests and diseases, ability to reduce transpiration losses in arid areas and minimum drain of nutritionist. Unlike aquatic plants, there is need to replace each harvest with a new crop as quickly as possible, the plants that have high sprouting potential following coppice cutting are preferable to those requiring replanting, using or breeding plants with leaves whose shape and orientation makes maximum of sunlight is another promising area to increase biomass production (i) ability to conserve water (ii) minimum drain of nutrient and ability to improve soil quality. The enormous diversity in the range of ecological conditions in India offers problems in afforestation and biomass production. For successful plantation trees must be raised in the nurseries later transplanted in the field. In order to achieve success in this newly developing challenging areas, steps need to intensify on resource survey and technological planning.

Technology assessment and refinement through institute village linked programmes require linear mechanism before their transfer. The programme approach involving various technology components relevant to the armours in varying agro-ecological zones for a perceptible change.

Discoveries of development of new technology by scientists usually in the form of principles, the bits and pieces of which are to be brought together for the development of appropriate technologies for the farmers. However commercial farm production systems with emphasis on off farm and on farm value additions? The entire programme seeks to develop multiple options for different target groups through a participatory mode.

For the mission mode programmes there would be a mission leader who may be the head of the Institution or a person charged with the responsibility of timely completion of the project. He will oversee the physical and financial programme of the mission and will be responsible for overall implementation of the programme in time bound manner. Wherever there are two or more centres the mission leader would also distinguish the function and release the funds including their monitoring and reporting for overseeing the progress at the lead and cooperating centres as well.

Principal investigator. Each research programme will be responsibility of a Principal Investigator located in the lead institution. He/She will be responsible for maintaining flow of technical activity from local level. Planning through implementation to monitoring and reporting. In addition the P.I. will ensure that emerging problems are recognised and address promptly referring to high project management, if participants in the programme maintain necessary and that all required financial records and bank accounts. He/She will be consolidated expenditure accounts and submit requests for replenishment of funds.

Priority for production system research. Agro-ecosystem and production systems for the potential contribution of improvements in the productivity stability sustainability of each production system are :

- National food security.
- Household food security.
- Environmental quality.
- Sustainability in the use of natural resources.
- Product diversification to enhance rural income.
- Welfare of tribal people, women and other disadvantaged group.
- Exploiting commercially viable technologies and enhance export potential.
- Improved mechanism to be developed. This will lead to adjustments in the choice of the programmes to be funded in future and the scale of funding, allocated to each from the national level.

Thematic areas. Research in each eco-system will revolve around the confronting problems :

- Natural resource management.
- Integrated pest management.
- Integrated plant nutrient management.
- Post harvest technology and value addition.
- Agro-biodiversity, waste management, water management.
- Socio-economic and policy research biotechnology.

The research technology and programmes would have the following common features :

- Technology assessment and refinement through institute village linkages.
- Planning through bottom-up and interactive approaches with the collaborating centres and formulation of individual proposals.
- Interdisciplinary inter-institutional research to improve production.
- On farm adaptive research to upgrade scientific research.

- HRD provision to upgrade scientific skills of the scientists, teachers to impart relevant education and out-reach, research and education.
- Global interaction for international institutions to develop long-term linkages.

Modes of Research

- Production system research : Under this system it will be location, specific, focus on local development needs, inter-institutional, interdisciplinary research synergies based on targeted production system in geographical areas.
- Cross cutting research : Concrete output achievable on strategic or production system, to provide data information to support research or critical gaps relevant to one or more production system for prioritising future investigations.
- Strategic research in terms of excellence.

Managing linkage mechanisms. Different types of mechanisms for strengthening link with technology users. Some, such as joint priority-setting exercises, demand intensive management, staff involvement, and funds. Others can be much more informal and simply help in the exchange of information. To build effective links, managers usually have to combine various types of mechanisms and apply them at different levels of the institutional hierarchy. They also need to maintain flexibility and be alert to setting up new mechanisms as institutional conditions and technologies change. Most of all, managers have to give the leadership and active management required for developing effective links.

Goal. Grants made in these areas allow the UGC to support special opportunities and circumstances that may fall outside traditional function boundaries.

Programme initiative funds. To a special funds for development applied for one time grant to develop programmes to conduct feasibility studies for functional literacy for HRD, curriculum for training programmes of pre-service, in-service and for collaboration with national and international graduates.

EDUCATION AND OUTREACH

Curricula for the Courses in collaboration with various National and International Universities

Ecology on wide scale for moving an individual molecule to a global ecosystem for human resource development at all levels in human ecology.

Technical Programmes

Ph.D in Human Ecology

Two year postgraduate programme in Human Ecology to prepare young men and women for managerial careers in developmental organisation.

M.Sc Biotechnology, Microbiology, Forestry, Human Ecology, Real Estate Education

One year (In-service, Pre-service) programme for development practitioners aims to providing opportunities for young officers in rural development and producers' organisations to strengthen their professional management skills.

B.Sc. Medical Biotechnology, Microbiology, Agriculture Forestry, Home Science, Biotechnology, Physiotherapy, Mass Communication and Tourism etc.

Three months General Management Programme provides opportunities to middle and senior level managers in rural producers, cooperatives for mid-career reorientation and training.

The six weeks Management Appraisal Programme for Voluntary Agencies provided exposure to the management principles to people already involved in implementing and managing development programmes.

Short duration Management Development Programmes, Seminars and Workshops bring a large number of developmental professionals to the department every year for specialised training for human resource management in the field of Human ecology.

Pre-service Courses Duration — 1 year

In collaboration with Texan A & M University, Texas, USA and College of Human Ecology, Michigan State University.

To equip all the students for graduate work for :

(a) Career in industry, research laboratory and Govt. service.

(b) For students desiring entry into patent law and environmental studies.

(c) Professional health science programmes and medical technology.

(d) Degree programme for fulfilling necessary pre-requisite of rural and underserved population through integrated population research and use of community focus health education and public service programme.

Modification of B.Ed. & M.Ed. Courses — 1 year

For quality education by providing foundation links relevant in the market place and leadership development opportunities in collaboration with Lowry Mays College and Graduate School of Business.

Curricula focus on teaching instructional programme and planning for Courses in education :

(a) Undergraduate certification and degree programme to prepare the students for role as elementary and secondary teachers.

(b) Graduate system in the department contribute to the professional educators knowledge based and prepare individuals for programme and instructional leadership roles and schools and other education setting including Colleges and University.

Education Programme for HRD

(a) Undergraduate degree in interdisciplinary technology.

(b) Masters and Doctoral degree include emphasis in adult and extension education, adult literacy, distance learning, industrial education, educational psychology, HRD and technology education.

(c) Degree in agriculture education — M.Ed., M.Sc.

(d) Allied health, teacher education and administration.

(e) Master of science and counselling.

Collaboration with NCERT and Department of Education, Govt. of India.

LOOKING FOR PARTNERSHIP ON HIGHER EDUCATION FOR REFLECTION

(To give big push to rural schemes for higher education)

Goal

International Centre for Human Ecology.

Sectors

Education for health, urban poverty, rural development and energy.

Measures

Infrastructure, HRD research on interdisciplinary curriculum, Establishment of nodal centre for planning, programming, monitoring and evaluation.

(HOME SCIENCE TO HUMAN ECOLOGY) PROPOSED SUBJECTWISE UPDATING THE CURRICULA FOR B.Sc. HOME SCIENCE COURSES OF STUDY

MODULE I	MODULE II	MODULE III	MODULE IV (OPTIONAL)
Agriculture and Environment (E.S.) Focus : Human Geography	Extension Education & Management (Community Resource Management)	Farming System Practices (F.N.)	Local Environment for Consumerism & Public Relations (R.M.)
Nutrition & Health Related Issues (For Balanced Nutrition)	Rural Sociology (Family Consumer Resources) (C.D.)	Production & Operation of Handicrafts Centre (T.C.)	Agribusiness, post harvest and processing, biotechnology, biodiversity & Environment and Non-conventional energy systems
Women & Child Development and their Welfare (C.D.)	Communication & Audio Visual Aids for Development & Advertising (R.M.)	Family Financial Management & Marginal Economics (R.M.)	Optional on farm, non-farm and off Farm sections (R.M. & MICRO BIO)

(*Contd.*)

MODULE I	MODULE II	MODULE III	MODULE IV (OPTIONAL)
Computer Basics : (a) Graphic Designing (T.C.) (b) Interior Decoration (R.M.)	Public Health Nutrition (Hygiene & Nutrition Diet) (F.N.)	Mgt. of Formal & Non-formal Entrepreneurial Development (R.M.)	Apparel & Accessories Designing (T.C.)
		Psychology (applied) Consulting self-esteem, coping stress (C.D.)	Creche (Child Development Principles) (C.D.)
			Religion & Cross Cultural Perceptive (C.D.)
			Industrial Psychology (C.D.) Criminal Psychology (C.D.)

Abbreviations used :
E.S. = Environmental Science,
C.D. = Child Development,
T.C. = Textile & Clothing,
R.M. = Resource Management and
F.N. = Food and Nutrition.

Objectives

- Financial mechanism by UGC for Nodal Centre.
- For providing opportunity to foreign universities and institutions to legitimate their operation in the country.
- To make mandatory for service, linkages support for renewal of projects, technical support, and regional connections.
- Proposal for the change the name — Centre for Adult

Continuing Education, Extension Work and Field Outreach to Centre for Human Ecology.

- To have command and total control over the functioning of the various projects.
- Regulate standards of education in institutions of country.
- To keep the project sites under 'watch' special watch for a particular period.
- Overseas utilisation of grants.
- To lay down norms, guidelines and mechanism and admission and for prescribing tuition and development fee to be charged by institutions.
- Foreign institutions for conferring the awards in India through regular distance mode or, by merely conducting their examinations they shall register themselves with the UGC.
- International Congress on Human Ecology for Women for Prioritisation of Research Agenda in Human Ecology in Collaboration...with to be Scheduled on...in New Delhi.

PLAN OF ACTION IN THE NEW MILLENNIUM

For undertaking the new research initiative and collaborative research and conducting training programmes for world-wide wisdom socially responsible and gender inclusive science and technology. Establishment of strong and influential networks of individuals and organisations that are engaged in research and grass-root activities that promotes gender equity in science and technology.

The beginning of the millennium is an appropriate time for reelection.

Objective of the Study

- To encourage research into all aspects of gender differentiation in science and technology education and employment.
- To foster gender equity in science and technology education.
- To facilitate the entry of women into employment in the

field of science and technology and their progress within such employment.

- To foster socially responsible and gender inclusion in science and technology.
- To provide a forum for dissemination and discussion of research findings and experiences of those in the field.
- To provide a support network for those working towards objectives outlined above.
- Issues of interactions between gender science and technology.

ISSUES OF INTERACTION

- Which scientific and technological advances would be liked to see in future?
- Which applications facilitated by the current state of science and technology are desirable?
- What demands will the answers to the above questions make upon the formal education sectors?
- How can science and technology education be constructed to embrace diversity of gender, ethnicity, class, religion, etc.
- How can informal and indigenous science and technology be acknowledged and integrated with formal science and technology?

In the first phase of the Congress on Human Ecology for Women, Department of Home Science, University of Delhi in congregation with IIT, Delhi and IARI has made a special effort to increase the relevance of higher education by making course more career oriented. To enhance empowerment in human resource with special emphasis on education the disciplines of Agribusiness, Biodiversity and Environment Biotechnology, Non-conventional energy system have been merged to make a unified field of Human Ecology.

Interdisciplinary knowledge and skills to deal with the problems through vigorous research, active extension education progamme will outreach the communities. This will help them to meet the challenges imposed by the new economic and social order and improve their quality of life.

In the new millennium India is well equipped, for its bio-industrial development. Today the philosophy of human ecology provides the context for developing and serving the individuals and their families from a human environment perspective towards building and sustaining a productive quality of life. The scientific expertise, infrastructural facilities and research needs to be strengthened by bonding the relationship between research and educational institutions and industry in a symbiotic mode of conversion of research findings into products processes and technologies for human resource development.

As we move to the 21st century, six hundred million students and point three one million teachers in India can play a major role in making an impact on the immediate society to which they belong. Adequate field experience will enhance knowledge with skills and develop appropriate attitudes towards equity and social justice with respect to gender, caste, clan and creed. The concepts and view points developed here will be helpful in framing educational strategies in higher education and research needs in the field of Human Ecology.

Under this Congress recommendation were made for future planning and action which are as follows :

- Expansion of research and outreach with people, resources and facilities.
- Postgraduate degree/diploma programmes for pre-service graduates.
- Other training programmes — short-term and doorstep (tailored programmes as per need).
- Research and training programme for wiring the curriculum of School and University.

COLLABORATIONS

1. Village Level

(i) Gram Panchayat.

(ii) Van Panchayat.

(iii) Fruit Utilisation Department.

(iv) Local NGOs.

2. National Level

(i) G.B. Pant University, Pantnagar.
(ii) Departments of Delhi University, Delhi.
(iii) Kumaon University, Nainital.
(iv) Forest Research Institute, Dehradun.
(v) Indian Council of Agriculture Research Centre, Almora.
(vi) G.B. Pant Institute of Himalayan Environment & Development.
(vii) National Academy of Agricultural Sciences, IARI Campus.

3. International Level

(i) Texas A & M University, Texas.
(ii) Texas Agriculture Extension Services, College Station, Texas.
(iii) Michigan State University, Michigan.
(iv) College of Human Ecology, Michigan State University, Michigan.
(v) The International Service for National Agriculture Research (ISNAR), The Hague, The Netherlands.
(vi) International Centre, The University of CALGARY, Canada.
(vii) ETC., Lusden, The Netherlands.

5

Action Mechanism for Research and Development

Globalization, a dominant force in the 20th century's last decade, is shaping a new era of interaction among nations, economies and people. It is increasing the contacts between people across national boundaries — in economy, in technology, in culture and in governance. But it is also fragmenting production process, labour markets, political entities and societies. So, while globalization has positive, innovative, dynamic aspects — it also has negative, disruptive, marginalising aspects.

The world is more prosperous, with average per capita incomes having more than tripled as global GDP increased ninefold, but these trends mask great unevenness — in the advances and in the new set backs. This year seeks to promote economic efficiency, generate growth and yield profits. But it misses out on the goals of equity, poverty eradication and enhanced human security. The rapid expansion of global markets the conditions for people, corporations and nations to compete globally, the urge to privatise and downsize public action in search of economic efficiently — creates an environment in which the needs of human development can be easily neglected, with spending subject to a fiscal squeeze. Reduced public spending weakness institutions of redistribution, leading to inequalities. And as individuals, compete in the global economy, they spend time in having their skills and working at a paid job — putting a time squeeze on caring activities. Care is also squeezed by reductions in pubic spending. And free market prices do not capture the full environmental costs of production and consumption, putting a squeeze on the natural environment.

Stronger policies for human development more investment of equip people for the globally competitive economy, and to participate in the global network society — are needed to promote human development. But they are also needed to make globalization work. Ultimately, people and nations will reject global integration and global interdependence if they do not gain from it and if it increases their vulnerability. Pressures will mount to retreat to isolationism in economic policy, culture and political priorities.

To pursue human development, globalization has to mean :

Ethics — less, not more violation of human rights and disregard of human values.

Development — less, not more poverty of countries and people.

Equity — less, not more poverty of countries and people.

Human security — less, not more vulnerability of countries and people.

Sustainability — less, not more depletion and degradation of the environment.

Despite the tremendous progress in the 20th century, the world today faces huge backlogs of deprivation and inequality that leave huge disparities within countries and regions.

Poverty is everywhere. Measured by the human poverty index (HPI-1), more than a quarter of the 4.5 billion people in developing countries still do not have some of life's most basic choices — survival beyond age 40, access to knowledge and minimum private and public services.

- Nearly 1.3 billion people do not have access to clean water.
- One in seven children of primary school age is out of school.
- About 840 million are malnourished.
- An estimated 1.3 billion people live on incomes of less than $ 1 (1987 PPP $) a day.

In industrial countries, too, human poverty and exclusion are hidden among statistics of success, revealing enormous disparities within countries. Measured by the human poverty index (HPI-2), one person in eight in the richest countries of the world is affected by some aspect of human poverty : long-term unemployment, a life shorter than 60 years, an income below the national poverty line or a lack of the literacy needed to cope in society.

The HPI disaggregated for a country's regions also shows wide disparities. In India, for example, the level of human poverty in the State of Bihar (54%) is more than twice that in Kerala.

Gender disparities remain large, too. In developing countries there are still 60% more illiterate women than illiterate men, and female enrolment at the primary level is still 6% lower than male enrolment. Disparities are starkest in the political and economic arena, with women nearly closed out of political life. Over the past decade dramatic events have changed the global political order, brought technological progress and shifted economic policies — events defining the character of globalization and greatly accelerating it. The end of the cold war unleashed a wave of global political integration. Information and communications technology has launched millions of global conversations. And the Marrakesh Agreement of 1994 changed the rules of global trade. All this in the wake of a global ideological shift.

GLOBAL INTEGRATION — RAPID BUT UNBALANCED

Global integration is proceeding at breakneck speed and with amazing reach. But the process is uneven and unbalanced, with uneven participation of countries and people in the expanding opportunities of globalization — in the global economy in global technology, in the global spread of cultures and in global governance. The new rules of globalization — and the players writing them — focus on integrating global markets, neglecting the needs of people that markets cannot meet. The process is concentrating power and marginalising the poor, both countries and people.

Today only 25 developing countries have access to private markets for bonds, commercial bank loans and portfolio equity. The rest are shut out by their lack of credit rating.

To sum up : the top fifth of the world's people in the richest countries enjoy 82% of the expanding export trade and 68% of foreign direct investment — the bottom fifth, barely more than 1%.

These trends reinforce economic stagnation and low human development. And they have further marginalised many developing countries from the most dynamic areas of global economic growth. The 1980s and 1990s have seen strong growth in the trade of manufactures, services and "knowledge goods". While some developing countries have made major advances, others have missed out entirely. Manufacturing exports should have been a step towards transforming their economies and creating more jobs. But only 33 countries managed to sustain 3% annual growth in GNP per capita during 1980-96. For 59 countries — mainly in Sub-Saharan Africa and Eastern Europe and the CIS-GNP per capita declined.

Economic integration is thus dividing developing and transition economies into those that are benefiting from global opportunities and those that are not. The uneven divide cuts across levels of income and human development — and across regions. Contrast China, Chile, Costa Rica, Mauritius and Poland with Cameroon, Niger, Venezuela and Russia.

Ironically, those left behind are deeply integrated in world trade. Sub-Saharan Africa has a higher export-to-GDP ratio (29% in the 1990s) than Latin America (15%). But Africa's exports are still mainly in primary commodities and foreign direct investment is concentrated in mineral extraction — so the region's apparent integration is actually a vulnerability to the whims of the primary commodity markets.

Countries are not the only major actors more and more it is multinational corporations that dominate global markets. Global employment opportunities may be opening for some, but they are closing for most others. The global market for high-skilled labour is now more integrated, with high mobility and standardized wages. But the market for unskilled labour is highly restricted by national barriers, even though it accounts for a larger share of international migration. Australia, Canada and the United States have programmes to attract skilled migrants, so the brain drain from developing countries. As many as 30,000 African Ph.Ds. live abroad, while the continent is left with only one scientist and engineer per 10,000 people.

GLOBAL CULTURE

Contacts between people and their cultures — their ideas, their values, their ways of life — have been growing and deepening in unprecedented ways. Television now reaches families everywhere. For many, the exposure to new cultures is exciting, even empowering. For others, it is disquieting, as they try to cope with a rapidly changing world.

As Mahatma Gandhi expressed so eloquently earlier in the century, "I do not want my house to be walled in on all sides and my windows to be stuffed. I want the cultures of all the lands to be blown about my house as freely as possible. But I refuse to be blown of my feet by any." Today's flow of culture and cultural products is heavily weighted in one direction — from rich countries to poor.

The rise of culture as an economic good has added to the identification of culture with commodities that can be sold and traded — crafts, tourism, music, books, films. Although the spread of ideas and images enriches the world, there is a risk of reducing cultural concerns to the vehicles for this trade in cultural goods are the new technologies. Satellite communications technology from the mid-1980s gave rise to a powerful new medium with a global reach and to such global media networks as CNN. The number of television sets per 1,000 people worldwide almost doubled between 1980 and 1995, from 121 to 235. The 1990s have seen a boom in multimedia industries, with sales of the world's largest 50 multimedia companies reaching $ 1.10. The development of the Internet is also spreading culture around the world, over an expanded telecommunications infrastructure of fiber optics and parabolic antennas.

GLOBAL GOVERNANCE

Governance is not government — it is the framework of rules, institutions and practices that set limits on the behaviour of individuals, organisations and companies. In today's integrating world there is clear need for global governance for the good of society, economy and environment. And a form of global governance is indeed emerging — but the imbalances in the process are cause for concern.

Intergovernmental policy-making in today's global economy is in the hands of the major industrial powers and the international institutions they control — the World Bank, the International Monetary Fund, the Bank for International Settlements. Their rule-making may create a secure environment for open markets, but there are no countervailing rules to protect human rights and promote human development. And developing countries, with about 80% of the world's people but less than a fifth of global GDP have little influence.

The rapidly increasing multilateral agreements — the new rules — are highly binding on national governments and constrain domestic policy choices, including those critical for human development. They drive a convergence of policies in a world of enormous diversity in conditions — economic, social, ecological. For example, most development countries previously exempted agriculture, medicines and other products from national patent laws, but with the passage of the agreement on Trade-Related Aspects of Intellectual Property Rights (TRIPs), almost all knowledge-based production is now subject to tight intellectual property protection, unified internationally. Further, the TRIPs agreement is unbalanced : it provides an enabling environment for multinationals, tightening their dominant ownership of technology, impeding and increasing the cost of transfer to developing countries.

These new rules and institutions advance global markets. But there has been much less progress in strengthening rules and institutions to promote universal ethics and norms — especially human rights to promote human development and to empower poor people and poor countries. Fortunately, two important forces of social governance are gaining strength.

Ironically, more attention has gone to enforcing labour and environmental standards in expanding free trade, using strong trade sanctions to punish countries that violate them. The Multilateral Agreement on Investment was being developed in the OECD to provide a predictable market for multinationals, protecting their rights. But no consideration went to their responsibilities to people — their responsibilities to limit their behaviour, to bind their obligation to respect human rights and to promote the development interests of the communities they touch.

Social Fragmentation — Reversals in Progress and Threats to Human Security

Uneven globalization is bringing not only integration but also fragmentation — dividing communities, nations and regions into those that are integrated and those that are excluded.

Social tensions and conflicts are ignited when there are extremes of inequality between the marginal and the powerful. Indonesia shows what can happen when an economic crisis sets off latent social tensions between ethnic groups — or between the rich and poor.

Recent research on complex humanitarian emergencies concluded that "horizontal inequalities" between groups — whether ethnic, religious or social — are the major cause of the current wave of civil conflicts. Inequalities — and in incomes but in political participation (in parliaments, cabinets, armies and local governments), in economic assets (in land, human capital and communal resources) and in social conditions (in education, housing and employment).

The shrinking of time and space is creating new threats to human security. The fast-changing world presents many risks of sudden disruptions in the patterns of daily life — in jobs and livelihoods, in health and personal safety, in the social and cultural cohesion of communities. Threats to human security can now speed their way around the world — the collapse of financial markets, HIV/AIDS, global warning, global crime. Global threats are increasing, outgrowing national abilities to tackle them, and outpacing international responses.

JOB AND INCOME INSECURITY

In both poor countries and rich, dislocations from economic and corporate restructuring and dismantled social protection have meant heavy job losses and worsening employment conditions. Jobs and incomes have become more precarious. The pressures of global competition have led countries and employers to adopt more flexible labour policies, and work arrangements with no long-term commitment between employer and employee are on the rise.

GLOBAL CRIME

Globalization opens many opportunities for crime, and crime is rapidly becoming global, outpacing international cooperation to fight it. There are now 200 million drug users, threatening neighbourhoods around the world. In the past decade the production of opium more than tripled and that of coca leaf more than doubled. In Belarus drug-related crimes increased from 4 per 100,000 people to 28 in 1990—97, in Estonia from 1.4 per 100,000 to almost 8. The illegal drug trade in 1995 was estimated at $ 400 billion, about 8% of world trade, more than the share of iron and steel or of motor vehicles, and roughly the same as textiles (7.5%) and gas and oil (8.6%).

Illegal trafficking in weapons is a growing business — destabilising societies and governments, arming in Africa and Eastern Europe. Light weapons have the most immediate impact on people's lives. Used in every conflict around the world, they have caused 90% of war casualties since 1945. In El Salvador the homicide rate increased 36% after the end of the civil war. In South Africa machine guns pouring in from Angola and Mozambique are being used in more and more crimes. In Albania there were five times as many murders in 1997 as in 1996, a rise attributed to the illegal arming of civilians.

Another thriving industry is the illegal trafficking in women and girls for sexual exploitation, a form of slavery, and an inconceivable violation of human rights. In Western Europe alone, about 500,000 women and girls from developing and transition economies are entrapped in this slave trade each year. Women lose not only their freedom, but their dignity and often their health. If they return to their homes, they are often rejected by their families and communities.

At the heart of all this is the growing power and influence of organised crime syndicates, estimated to gross $ 1.5 trillion a year — a major economic power rivaling multinational corporations. The sheer concentration of their power and money criminalises business, politics and government. Look at the Six Triads in China, the Medcllion and Cali cartels in

Colombia, the Mafia in Italy, the Yakuza in Japan, the Juarez, Tijuana and Gulf cartels in Mexico, the Cosa Nostra in the United States and the organisations in Nigeria, Russia and South Africa. All have operations extending beyond national borders, and they are now developing strategic alliances linked in a global network, reaping the benefits of globalization.

THE SPREAD OF HIV/AIDS

Global travel spreads more than ideas. The latest estimates by UNAIDS and the World Health Organisation show that more than 33 million people were living with HIV/AIDS at the end of 1998. The spread of the virus continues unabated, with 11 men, women and children becoming infected each minute — about 6 million in 1998. AIDS causes 2.5 million deaths a year, more than twice as many as the 1 million deaths from malaria. Yet some experts say that we are only a tenth of the way into the epidemic.

AIDS is now a poor people's epidemic, with 95% of all HIV-infected people in developing countries. HIV/AIDS has taken a heavy toll on the life expectancy built up over the past three decades. A loss of 17 years in life expectancy built up over the past three decades. A loss of 17 years in life expectancy is projected for the nine countries in Africa with an HIV prevalence of 10% or more — Botswana, Kenya, Malawi, Mozambique, Namibia, Rwanda, South Africa, Zambia and Zimbabwe — down to 47 years by 2010, back to the life expectancy of the 1960s.

HIV is also spreading fast in areas thought to be relatively free of the virus — in China and even in the vast rural areas of India, where some studies show higher prevalence rates than in urban areas. Eastern Europe and the CIS had appeared to be spared the worst in the early 1990s, but new surveys show stupendous increases in Belarus, Moldova, Russia and Ukraine. There, too, HIV/AIDS is often associated with poverty, spreading among marginalised people, especially through drug use.

CIVIL CONFLICT, GLOBAL UNREST

Civil conflicts have been flaring for decades. What's new today is the complex interaction of interests, the blurred line between conflict and business. Defence is becoming privatised, and international private military firms are proliferating. In some countries mercenaries often sell their services for mining and energy concessions and set up affiliates in air transport, road building and trading. And more and more, the clients of mercenaries are multinational corporations seeking to protect their mining interests in conflict-prone countries.

Executive Outcomes, Sandline International and Military Professional Resources Incorporated offer military services and training to governments and large corporations and have been particularly active in Africa. The Mobutu government in its final days spent some $ 50 million in desperate attempt to stay in power in the Democratic Republic of the Congo. The rise of military companies is linked to the post-cold war power vacuum. Major powers are less inclined to intervene militarily, especially in low-level conflicts.

Accountable only to those who pay, such businesses are hard to regulate. So far, domestic and international laws seeking to limit mercenaries' operations have been ineffective. The annual reports of the UN Human Rights Special Rapporteur on Mercenaries have regularly urged governments to develop legislation that bans the use of mercenaries in their territories.

ENVIRONMENTAL DEGRADATION — A SILENT EMERGENCY

Environmental degradation is a global problem that surpasses the scope of national governments. Globalization can improve prospects for environmental management — through the spread of environment — friendly technologies, standards and pressures by consumers and activists. It can also add pressures for environmental exploitation — export-led demand for paper leading to deforestation, and demand for fish leading to overfishing.

Environmental degradation is a chronic and "silent emergency" that threatens the livelihoods of some of the poorest people of the world. Scientists predict a steady rise in global temperatures and sea levels, inundating as much as 17% of the land area in Bangladesh, 12% in Egypt and almost all of the Maldives. Renewable resources are being depleted rapidly and unsustainably : fish stocks are three quarters of what they once were. Water availability today is 69% of 1970 levels, as is forest coverage. All this threatens the economic security, food security and health security of the world's poorest people.

6

Agricultural Research and Education Management

- Community Approach : A Perspective

INTRODUCTION

What of thee I dig out, let that quickly grow ever. Let me not hit thy vitals or thy heart.

The Atharva Veda

The hymen about conceptualises thee as natures are, refers it to man warning of him of the relationship he must maintain to live in harmony with nature.

Ecology

Relationship between environment, body and life is the theme of ecology. This was very well-known to the Indian sages of the past, though the evidence of systematic development of ecology as a science is lacking. Modern science and technology have come to realise the importance of this discipline, and lately there have been significant advance in this direction, both as at pragmatic as well as academic level.

Thus *ecology* is a developing science and still more, is the awareness of the value of the recent environmental crises that has stimulated the politicians, administrators and the public alike in the understanding of natural ecological perspective. Hence suitable literature has turned to the recent advances in the discipline, essential for the dissemination of this knowledge.

Ecology is having a new look concerned with the system that supports man. It emphasises on the study of the behavioural aspects coupled with the inter-relationship of the biotic and a biotic environmental components. The centre of the study is organism not the environment. An integrated approach to the study of individuals, population, community, plants, animals and man is provided through the study of ecology. The materials for ecological studies are the organisms and the environment in which they live. There are two distinct components to the environment, the physical environment (comprising such things such as temperature, water availability, wind speed, soil acidity) and biotic environment, which comprises any influence on an organism that is exerted by other organisms, including competition, predation, parasitism and cooperation.

Within ecology there are number of fields, either focusing specific areas of interest or using particular approaches to address ecological problems. For example *behavioural ecology* is concerned with explaining the pattern of behaviour in animals. Physiolgical ecology explores the physiology of individual and considers the consequences on function and behaviour. There is a particular emphasis on the impact of evolution on current pattern in the focus of evolutionary ecology. The use of molecular biology to directly tackle ecological problems. The domains of population ecology and community ecology are not restricted to natural system. Understanding both the human effects on nature and the ecology of artificial environments (such as crop fields) are important areas of study.

Ecology can be considered on a wide scale moving from an individual molecule to the entire global ecosystem. However, four identifiable sub-division of scale are of particular interest : (i) Individuals, (ii) Population, (iii) Community and (iv) Ecosystem.

At each scale the subject of interest is to ecological change. At the individual level the response of individual to their environment (*biotic* & *abiotic*) is the key issue, while at the level of population of single species, the determinants of abundance and population fluctuation dominates. Communities are the mixtures of population of different species found in a defined area.

Man is an important element of the environmental complex. His activities, growing population, industrial and urban settlements, social, religious and economic set-ups posing newer ecological problems. He plays an important role. Understanding the ecological role, the anthropologists, sociologists, psychologists, economists and demographers supply data to the persons working in human ecology. Growing activities of man and his increasing demands on the biotic and abiotic components of the environment has brought forth a change in his ecological philosophy. This is now directing his attention more on the study of habitat or ecological systems as a whole than on their components. The perspective now is the studies of the organism to those of habitat or the ecological systems or the ecosystems. The ecosm is considered the functional units of the biosphere.

Man is the master consumer, uses the biotic and abiotic resources of biosphere and at places has created ecological crises. Keeping in view the production potential of the different types of ecosm, we can plan our activities and resource management based on structural and functional attributes of the ecosms or ecosystems. Each is characterised by functional differences and production potentialities. Their studies involve team work and multidisciplinary approach, to evaluate their production potential for the human welfare.

In order to live let live man has to be educated to a proper understanding of environment and management of natural resources through proper conservation practices. Ecological principles are being applied in forest management, grassland management in many developing countries. Ecology in every day science and man as the master consumer of natural resources must be properly educated to understand the impact of changing environment and depleting resources on his future progress.

Data obtained by ecological investigators are now being computerised and models are being prepared by various workers. This has resulted in the emergence of "*Statistical ecology and system ecology having great future*".

Commensurate with new challenges and opportunities both at national and global level has reoriented its mandate to include human resource development and technology assessment and refinement as new paradigms for holistic advancement in agriculture and allied subjects while ensuring optimal use of natural resources in partnership with all stakeholders especially, farmers, NGOs, private sector, urban slum community so as to achieve the ultimate goal development.

To achieve the twin natural objectives of poverty alleviation and welfare of the disadvantaged and weaker section of the society, it has become imperative to increase agricultural productivity, profitability, and sustainability through judicious and more integrated use of our natural resources.

The agrarian structure in the country is rapidly changing under the force of population growth, industrialisation, urbanisation and technological development especially in the field of agriculture. The land use pattern is against the natural balance of ecosystem.

In spite of institutional reforms, small and marginal farmers are not able to survive and there is need to have a floor of land holding as we have a ceiling. The small and marginal farmers are at the edge of conversion into agricultural labourers.

Agriculture will continue to be main occupation of majority of rural population for long time to come. The growing population demands a lot more efforts in the field of research education and extension. In view of large number of productive technologies available everywhere in the country, it is high time that transfer of technology gets priority over research and education. Rural household and farming systems have become a prime focus of development assistance in recent years. Both in applied research and in project work a farm family perspective is being adopted to achieve successful agricultural development.

The farm is no longer viewed solely as a productive unit. The focus has shifted to the rural household in its relation to its socio-economic and cultural environment. The rural family is seen as the central decision-making unit for the utilisation of existing development potentials and application of innovations. Simultaneously the rural household accounts for social stability in the society and for securing the material and non-material essential needs of its individual members.

Existing development efforts including research and action are after following "top down" approaches in developing planning and implementation. These conventional approaches often do not address the constraints and development potential of the farm household system and rural communities in comprehensive way. FSD has been developed to overcome these shortcomings and should therefore, be seen as complementary to existing approaches the farm household is the principal system and focus of FSD.

The farm household systems are rural households consisting of the three basic subsystems, which are closely interlinked and interactive.

1. The household as the decision-making point.
2. The farm and its crops and livestock activities.
3. The off-farm component.

In FSD the farm household system is viewed as an integrated part of the prevailing farming systems in the rural communities.

Rural communities are the larger system, comprising many farm, non-farm households. The non-farm household systems comprise many different sub-systems which are also closely interlinked and interactive. Traders, retailers, social groups, government institutions, leadership structures politics, religion all belong to non-farm household systems. They are complex and have their own organisation and management. Farm household systems often participate and interact with several of these systems.

To develop farm household system and rural communities, it is necessary to clearly understand the objectives and functioning of farm household system in relation to their physical, socio-economic, and policy environment. FSD, therefore, works closely with rural people and community.

The interactions between farm households, other members of the community, the government and the interests of other socio-economic groups, need to be analysed and understood.

Constraints and development potential of the farm household system must be identified in a participatory approach.

The analysis and constraints and potentials include external factors, such as price, policies, and by laws, the organisation and management of agricultural support services, input supply, etc., because they affect the functioning of the farm household system, directly or indirectly in many different ways.

FARMING SYSTEM DEVELOPMENT (FSD)

Potential intervention packages identified with FSD approach are listed and their effects are checked on the whole farm household system, and on other systems, such as non-farm household systems and the environment, wherever possible. FSD stresses communication to policy makers, support services, research and extension about the action for development.

FSD is, therefore, a continuous, dynamic and interactive learning process based *on analysis and testing, monitoring, evaluation and feedback.*

Development must ensure continuous benefits to the rural people. Resources and environment need to preserved for future generations. This approach of FSD can be initiated through training courses. This has the advantage that a number of people learn about the approach and can apply FSD principles in their daily work. Equally they can motivate team members to think along the same lines. There also can be "full" FSD applications, the steps are identical with the modules of training course.

The steps in FSD :

Module I — Introduction to farming system development.

Module II — The concept and approach to farming systems development.

Module III — Preparatory arrangements when starting FSD.

Module IV — Farming Systems zoning.

Module V — Exploratory diagnosis and analysis.

Module VI — Verification.

Module VII — Identifying constraints and potential improvement.

Module VIII — Evaluation of potential involvement.

Module IX — On-farm testing.

Module X — Planning interventions and implementing how to prepare for implementing FSD.

(This module considers how to design development programmes, how to prepare for implementing FSD, and how to coordinate, planners, researchers, extension support, services and farmers for and FSD Thrust.)

THE TECHNOLOGY TRIANGLE

It has been realised that developing countries concern about persistent rural poverty, malnutrition, unsustainable agricultural production, lack of food sufficiency, and swelling urban populations. In response many governments have undertaken major initiatives to help poor farmers households improve their living standards and productive capacity. These farmers make up not only the majority of producers in many developing countries, but also a major share of world's poorest and most vulnerable poor.

Most research systems in developing countries were organised to serve commercial farmers operating in more favourable and homogeneous agro-ecological conditions than those in response to poor farming context. Traditionally, resource poor farmers have lacked the power and organisational ability to exert pressure on the research systems to meet their needs. Their access to research information is restricted; their ability to articulate, their need is poor; their capacity to tolerate risk is limited; and pressing concerns of their daily existence make it difficult for them to focus on long-term technological change. The challenges facing research institutions, therefore, is to develop a greater capacity to facilitate effective interaction between researchers, technology transfer workers and resource poor farmers.

There is no single recipe for strengthening links. The policy and institutional context determines the type of strategies and mechanism as a manager can use to develop effective links. Key contextual factors influencing links are agricultural development, and research policies, the resource situation and organisational structure of the institution involved, and technical issues such as existing knowledge base, the inventory of available technologies, and the diversity of agro-ecological conditions and production systems.

The policy context in which an institution operates is shaped mainly by external pressure from national policy makers, foreign donors, the private sector and in some cases farmers' organisations. These pressures can stimulate institutions to improve performance, build stronger links and address the needs of resource poor farmers. However, they do have limitations. They tend to focus on short-term goals and often overestimate the capacity of local institutions to meet new demands.

ORGANISATIONAL FACTOR

When developing linkage strategies, research managers need to consider organisational structure both an entity within which they have to manoeuvre and as a variable they can manipulate. The size of an institution is a critical factor affecting links. Small institutions, for example, can benefit more from informal links among staff. Yet they often face severe resource constraints in terms of staff and funds, and have trouble in sustaining even most basic linkage mechanisms.

A key factor, particularly for larger, more complex institutions is how task and responsibility are divided among organisational units merging research and technology transfer or on-farm and on-station research into one department and institution is often proposed as a solution to linkages problems. This can be successful but only under certain conditions. The group must share a common focus, such as single commodity; they must have same level of commitment to working together towards a common goal, and the institutions should not be too large. Although separating groups into different unit set up organisational barriers, it has some important advantages. It encourages specialisation and development expertise, and often permits closer supervision and leadership which approach is more appropriate depends on the context.

Three types of structural mechanisms are commonly used to establish links between separate, but interdependent units; direct supervision by a common manager; coordination units or positions, or permanent committees made up of representatives from relevant groups.

MECHANISM

In addition to structural linkages, managers can use four basic types of mechanisms to strengthen links, joint planning and review processes; collaborative professional activities, resource allocation producers; and communication devices. Various types of mechanisms are appropriate for different kind of linkage problems, have distinct managerial and resource requirements, and vary in case of implementation. Moreover, different types of technologies require different type of linkage mechanism.

To build effective links with technology users, managers need to use a combination of various mechanisms and apply them at different levels of the institutional hierarchy. All linkage mechanisms cost time and money. Managers need to choose them carefully, apply them frugally and adapt or replace them with new mechanisms as technologies and institutional conditions change. Above all, managers (Principal Investigators) need to provide leadership and "hands on" management in developing links.

MANAGEMENT ISSUES

No linkage mechanism can succeed unless working on research stations, on farms, and in technology transfer agencies are motivated to collaborate. The challenges for managers is to get these staff with their differing backgrounds, skills, aspirations and responsibilities, to work together and to recognise that they depend on one another to reach a common goal.

Perceived status differences between researchers, technology transfer aspect workers, or even on farm workers and in technology transfer agencies, often impede collaboration. Status problems have no simple solutions. Managers have three basic options; reduce the difference through training or increase professionalisation; accept the difference, but work to minimize their negative impact, even in difficult situation. Avoid status problems by finding alternative partners, or building up their own capacity in technology transfer. Managers make the difference between strong and weak links. Active management means providing leadership maintaining flexibility, and responsiveness, and having ability to manage conflict. They promote mutual respect and a feeling of interdependence between the groups. And they make sure that staff not just the managers feel that they benefit personally from collaboration.

In view of large number of productive technologies available everywhere in the country. It is high time that technology transfer gets priority over research and education. There are five extension systems; the first line extension education role is basically the institutional role in transfer of technology. This role is extremely important as a catalytic force, so that the time lag between technology generation and technology adaptation is reduced. The project/institution should focus to the following objectives :

1. Promptly demonstrating the latest agricultural technologies to the farmer as well as the extension workers of the station with the view to reduce the time lag between technology generated and its adoption.
2. Testing and verifying the technologies in the socio-economic conditions of the farmers and identifying the constraints.
3. Getting the first hand scientific feedback by the scientists in order to keep abreast with the performance of technologies and the farming problems, and reorient their research, education and training programme.
4. Providing training and communication supports to the state department of agriculture of NGOs.
5. Based on field work and experience, develop extension modules to be adopted by general extension for large scale multiplication.
6. Promoting research in extension and studies, including comparative studies of extension systems in different parts of the world.

The goal of every research and extension institution in the world is to provide relevant technology, transfer it to farmers in an efficient manner, and thus have a sizable impact on agricultural production and farmer's welfare. To reach this goal obviously requires links between the institutions and outside world (national — international agencies and policy maker); within the institution, among researchers, and between the institution and its clients, and the farmers. But these links do not develop spontaneously. They are costly to operate and take time and creativity to manage. When we need so many linkage mechanisms that an organisation becomes excessively complex, we have to start thinking about organisations rather than creating more positions or units in the organisation.

Finally the goal of developing links to build bridges between researchers and extension workers, should not make us forget that specialisation and division of labour have contributed to growth throughout history. Sometimes a bit of competition between different actors, will lead to improved performance by each of them, as long as the social and institutional goals are clear for everyone.

HUMAN DEVELOPMENT

In 1990 the time had come for a broad approach to improving human well-being that would cover all aspects of human life for all people, in both high income and developing countries both now and in the future. It went far beyond narrowly defined economic development to cover the full flourishing of all human choices. It emphasised the need to put people — their aspirations and their capabilities — at the centre of the development effort. And the need to assert the unacceptability of any biases or discrimination, whether by class, gender, race, nationality, religion, community or generation, human development has arrived.

Human development is the process of enlarging people's choices — not just choices among different detergents, television channels or car models but the choices that are created by expanding human capabilities and functionings — what people do and can do in their lives. At all levels of development a few capabilities are essential for human development, without which many choices in life would not be available. These capabilities are to lead long the healthy lives, to be knowledgeable and to have access to the resources needed for a decent standard of living and these are reflected in the human development index. But many additional choices are valued by people. These include political, social, economic and cultural freedom, a sense of community, opportunities for being creative and productive and self-respect and human rights. Yet human development is more than just achieving these capabilities; it is also the process of pursuing them in a way that is equitable, participatory, productive and sustainable.

Choices will change over time and can, in principle, be infinite. Yet infinite choice without limits and constraints can become pointless and mindless. Choices have to be combined with allegiances, rights with duties, options with bonds, liberties with ligatures. Today we see a reaction against the extreme individualism of the free market approach towards what has come to be called communitarianism. The exact combination of individual and public action, of personal agencies and social institutions, will vary from time to time and from problem to problem. Institutional arrangements will be more important for achieving environmental sustainability, personal agency more important when it comes to the choice of household articles or marriage partners. But some complementarily will always be necessary.

Getting income is one of the options people would like to have. It is important but not at all important option. Human development includes the expansion of income and wealth, but it includes many other valued and valuable things as well.

For example, in investigating the priorities of poor people, one discovers that what matters most to them often differ from what outsiders assume. More income is only one of the things poor people desire. Adequate nutrition, safe water at hand, better medical services, more and better schooling for their children, cheap transport, adequate shelter, continuing employment and secure livelihood and productive, remunerating, satisfying jobs do not show up in higher income per head, at least not for some time.

There are often non-material benefits that are often more highly valued than material improvements. Some of these partake in the characteristics of rights, others in those of states of mind. Among these are good and safe working conditions, freedom to choose jobs and livelihoods, freedom of movement and speech, liberation from oppression, violence and exploitation, security from persecution and arbitrary arrest, a satisfying family life, the assertion of cultural and religious values, adequate leisure time and satisfying form of its use, a sense of purpose in life and work, the opportunity to join and actively participate in the activities of civil society and a sense of belonging to a community. These are often more highly valued than income, both in their own right and as a means to satisfying and productive work. They do not show up in higher income figures. No policy maker can guarantee the achievement of all, or even the majority, of these aspirations, but policies can create the opportunities for their fulfilment.

Global Proposals

Global proposals have been aimed at contributing to a new paradigm of sustainable human development — based on a new concept of human security, a new partnership of developed and developing countries, new forms of international co-operation and a new global compact.

Human Development as a National Tool

The geography, history, demography and ecology poses a unique development challenge to develop its economy, infrastructure facilities and raise the living standard of its people. These are always alternative choices and economy to choose which serve the interests of its people best. However, there are difficulties in selecting best alternative, especially more so when the development involves not a large country but a group of small section of the society or a village, city or a country as a whole.

The philosophy of Human ecology provides the context for developing and serving the individuals and their families a productive quality of life. To take advantage of growing opportunity, matching human resources has to be developed. In view of rapid technological advancement, information explosion, diverse condition and growing capabilities has made it imperative to be developed HRD skills people for development. Further increasing the agricultural production for development in isolation is not the only issue but we have to address the issues of sustainability, environmental and ecological imbalance, depleting natural resources and widening of regional and socio-economic inequity and have to maintain the biodiversity.

Interdisciplinary knowledge and skills to deal with the problems through vigorous research, active extension education programme will outreach the communities. This will help them to meet the challenges imposed by the new economic and social order and improve their quality of life.

Higher education provides people with an opportunity to reflect on the crucial social, economic, cultural, moral and spiritual issues facing humanity. It contributes to national development through dissemination of specialised knowledge and skills.

EDUCATION AND RESEARCH MANAGEMENT FOR COMMUNITY SERVICE

(Revamping of programs of rural development, which needs attention on investment in agriculture for extensive agricultural research and extension system, emphasis on entire farming system taking into account regional diversity in resource endowment, control on water and land degradation.)

Towards Sustainable Communities

Activities that the environment can sustain and that citizens want and can afford may be quite different from community to community rather than being a fixed thing. A sustainable community is continually adjusting to meet the social economic needs of its residents while preserving the environment's ability to support it. A sustainable community seeks a better quality of life for all its residents while main learning native's ability to function over time by minimizing waste, preventing pollution, promoting efficiency and developing local resources to revitalize the local economy. Decision-making in a sustainable community stems forms a rich civil life and shared information among community members.

A sustainable community resembles a living system in which human, natural, economic elements are interdependent drains strength from each other. For the first time in history nearly half of the world's people now live in urban areas characterised by human structures and activities. The way our communities develop will largely determine our success of failure in overcoming environmental challenges and achieve sustainable development. Cities, towns and villages can provide enormous, untapped opportunities; they must and can pioneer new approaches to sustainable development and community management.

Mandate and Criteria

Awareness on environmental issues is a key element of any nations' ability to meet the challenges of conserving nature and protecting the environment. By motivating and channeling the resourcefulness of India's young people, and understanding of the interrelationships of all aspects of the environment can be encouraged throughout society. This will help to sensitise key target groups, specially children, students, teachers for awareness and action orientation. This will help in promoting nature conservation and environmental protection and develop community initiatives that encourage and enable a wide variety of interest groups to the contribute to the conservation of nature.

The prime focus of the project is to develop natural resources and improve environmental managerial capacity within government, non-government/community organisations, and with the private sector as the basis for addressing specific natural resource and environmental problems in water and energy sources in India for capacity building objectives :

Institutional Development; Conservation and Restoration; Policy and National Dialogue which will have an impact on national and state policies and related practices; Technology adoption and dissemination supportive of applications that demonstrate : (a) Community participation in project implementation, (b) direct environmental benefits through physical improvement, (c) a promising or innovative approach for integrated resource management.

Collaboration for Capacity Building

To restore degraded land and improve their living conditions for primary farm forestry. Indian Farm Forestry Development Cooperative. (IFFDC) Indian Farm Fertilizers Cooperative Limited. (IFFCO)

For promotion agro-forestry towards promotion of mixed cropping and generation of additional income through tree plantation. Nagaland Environment Protection and Economic Development through People's Action (NEPED).

Demonstration in improving environmental quality in stabilising crop production through management of the land, water resources to improve natural resource conservation, management and socio-economic conditions. For environment friendly income generating activities and promotion of renewable energy for domestic use. Environmental Improvement in Rainfed Areas. (EIRA) Implementing Organisation : Indian Potash Limited. (IPL)

For balance ecosystem and grass-root action for utilisation of common property resource and fuel, fodder and drinking water. Water Resource Development and Energy Conservation for Sustainable Management of the Environment. BAIF Development Research Foundation.

For promotion of sustainable community base approaches to be process of halting and reversing environmental degradation. Management of Environment Resources by Communities. (MERC) Aga Khan Foundation (AKF), India.

To check movement of sand and fluvial deposits to arrest and reverse the trend of environmental degradation and build local community capacity for the catchment area. Integrated Watershed Management and Water Storage in Pushkar Lake Area. Department of Watershed Development and Soil Conservation. (DWDSC)

To demonstrate and promote the feasibility of using waste material such as flash and locally biomass to regenerate degraded lands by plantation in wastelands. Land Restoration Through Waste Management of India. Indian Institute of Technology, Kharagpur.

To improve vegetative cover, increase community participation and enhance crop yield, leading to improve ecological balance and income generating opportunities for environment regeneration.

Project for Environment Regeneration.
MYRIDA

To demonstrate application of integrated low-cost modern and traditional technologies, eco-restoration through appropriate catchment area treatment measures, strengthen participatory management in local context, and capacity building of implementing groups and beneficiaries.

Sustainable Drinking Water Supply in the Rural Watersheds of the Humid Tropics in Kerala.
Malanadu Development Society (MDS), Kerala.

For the demonstration and feasibility of renewable energy technologies for local communities to take up increase use of renewable energy systems and energy conserving technologies for meeting their requirement of electricity and for taking up marketing, servicing and financial activities for disseminating renewable energy systems.

Implementation of Renewable Energy Technologies in Rural India through Non-Governmental Organisations (NGOs).
Tata Energy Research Institute, New Delhi.

To develop implement, demonstrate and disseminate a bio-energy package to meet comprehensively the rural energy needs in a sustainable and participatory manner to promote rural development for increase spread of bio-energy packages development for the installation of gasifier energy system, cooking gas, lightning and water supply for domestic use and irrigation.

Bio-energy for Sustainable Rural Development in Tumkur District, Karnataka.

Karnataka State Council for Science and Technology (KSCT).

Research Priorities for the Cause of Knowledge

- System based, location base adaptive research and natural resource management for commodity and discipline dominated programs.
- Enhance efforts to improve links between researches and extensionist, with more active involvement of Universities in fostering links with extension.
- Greater use of modern information technology and computerisation of scientific data access and dissemination, networking research management and monitoring.
- Efforts to attract private funding, where appropriate to technology generation, assessment and transfer.
- Piloting of a new institution village linkages programs to increase exposure of applied and adaptive researches to the system at which their investigations are aimed.

Development and management programmes. Swiss development co-operation to advice in carrying out reform programs in education and management for training effectiveness may be required.

Mission. Strengthening and empowering individuals and families to interact effectively with their environments.

Values. Dynamic ecological and global perspectives of people, contexts events, knowledge that improves daily lives of current and future generations.

ACTION MECHANISM

The key subjects namely Advocacy, Education, Women Employment, Health and Nutrition, Women in special needs, communication etc. to build up capability, to able to ensure that the delivery arrangements for programmes should be efficient and they should reach those for whom they are meant. The voluntary organisations have emerged as most viable institutional framework to serve as catalysts for women's development. It would be very important for the government to invest funds in the women's movement and provide political support to it.

Our networking will not only liberate; built blossom the long suppressed *Stree Shakti* of our people. In education new educational methods must be adopted to ensure that the girl child should be given proper education, as they remain illiterate especially in villages and slum dwelling women. No proper attention was given to adult literacy therefore, there should be functional literacy programme for women instead of Adult Literacy programmes. Empowerment should be generated among the women. It cannot be transferred from outside. Mobilisation of women folk, mobilisation of resources and participation of women in policy formulation in any perspective of society should not only involve the Government and Non-Government sectors but also equally involve the men. This should be taken as holistic approach. Empowerment can only be done economic upliftment. Though it is criticised that 33% participation for women in Panchayati Raj is ineffective as they have not acquired proper skill but in due course of next 5 years they will definitely acquire it. Collaboration with NGOs to bring economic development and empowerment of women by individual, behaviour and counselling or NGOs. It should also include some men in their managing body. It was stressed that some public inquiry should be conducted by NCW and it should monitor the land distribution agricultural wages. Existing policy implementation labour laws etc.

SHG Group building is very important platform in the sense that the community development activities is better fared. There is total lack of organised economic activity on the part of slum women. It was felt that if these women are empowered they have to be provided an infrastructure and awareness regarding health hygiene and other housekeeping activities to work in houses. Empowerment of girls to deal with their own common ailments, learning of Indian system of medicines should be promoted for housewives; political interference in various projects such as primary health projects should be stopped. In the absence if identifying the mother unit the NGOs are unable to work.

All the NGOs should be asked to promote awareness about the importance of nutrition and hygiene so that a healthy life could be promoted. The nutrition needs should be linked with some income generating activities so that the nutritional needs could be properly implemented. Vocational training should be provided to the inmates of every home and the effort should be that these institutions do not become permanent homes for the inmates. More than 80% workers among women are engaged in agricultural activities. Women work is much more as they participate in farming and animal husbandry work besides attending to household duties. It was realised that there is need for transfer of technology to women so as to improve production and efficiency and reduce drudgery in occupation of farming women.

The Chairman Dr. K.L. Chadha and Dr. P.L. Gautam felt the need for the constitution of expert committee to examine the related issues for especially in the light of centres, departments, universities and women study centers to take up, functional literacy infrastructure for economic empowerment for women through various strategies in the light of school curriculum and adult education programmes in slums for the women who have migrated from tribal areas to cities and working as domestic help, should be organised to learn through training for functional literacy to work with confidence. Secondary education has its importance in the national development as it provides large skilled manpower for the economy as prepare for higher, technical and professional education. It is the stage where a proper understanding of work ethos and values of humane and composite culture is provided to future citizens of the country.

There is a great demand for expansion of higher education an account of increasing expansion of primary and secondary education and rising social expectation. Consequently emphasis has been on the removal of disparities and to equalise educational opportunities attending specific needs to those who has been on the removal of disparities and to equalise educational opportunities attending to specific needs to those who have been equality so far.

Educational development is participatory process. Education was incorporated in the list to facilitate evolution of all Indian policies in the field of education. Government has assumed a larger responsibility to look after the integrated character of education, to look after international aspects of education, culture and human resource development; in the general to promote excellence in the sectors of education throughout the country. Having regard to the size of population, diversity and economic frailties the challenges are enormous. This realisation has resulted in steadily rising levels of budgetary allocations, and resource transfer for education to the states.

Elementary education. Universal access, universal relation and universal achievement are broad parameters to achieve the goal of universalisation of elementary education is yet to be achieved. Considering the magnitude and the complexity of the problem, meticulously formulated strategies to have been adopted in the form of operation black board (OB), Non-Formal Education (NFE), Teacher Education (TE), district primary education program to address the specific problem areas, gender and regional disparities. Special focus has been laid on the access to education for millions of girls and working children, who cannot participate in the school system on account of socio-economic compulsions. These segments largely constitute girls SCs/STs and other deprived sections of the society.

Although special programmes have been taken up to improve upon specific deficiencies in the educational system, no strategies can succeed in achieving UEE without specifically addressing its gender and regional dimensions.

Adult education. The national literacy mission represent a national and political commitment on the need to harness all social forces to achieve the objective of eradicating illiteracy of the people. It has made the country to set a target to make 100 million persons in the age group of 15 to 35 years, functionally literate in the future.

The total literacy comparing in the principal strategy adopted by NLM for eradication of illiteracy in the country. The literacy campaigns are area specific, time bound, and delivery through voluntarism, cost effective and outcome oriented. The LM as undertaken a number of new initiatives in adult education to ensure mass participation and decentralising the financial and administrative powers to state literacy mission authority.

Secondary education. Secondary education which serves a bridge between primary and higher education serves a step towards preparation for higher and professional education.

Socio-economic imperatives and initiatives taken during the Eighth Plan to provide learning facilities to school going children upto elementary stage has necessitate the perspective for secondary education sector in the Ninth Plan for harnessing technical innovations for improved teaching/learning and quality upgradation. The key issues in secondary education apart from access are quality, relevance, modernisation and diversification. Emphasis has been laid on programme : value education, population education, and various newly emerging human rights, adolescence education. Greater diversification and flexibility, proper education materials, capacity building in the schools to gradually eliminate the domination of public examinations are some of the aspects of modernisation of curriculum. The NCERT continues to play a nodal role in evaluation of textbooks from stand point of national integration and incorporation core elements outlined by NEP.

The National Council of Educational Research and Training continues to provide major resources support to school education in the form of various experimental innovative programmes; to diversify and make qualitative improvements in secondary education and also reducing the burden of school bags.

INSTITUTIONALISATION

Government of India has been devoting much attention to science and technology programmes ever since its independence. The problems of integrating tribal women in the development process are now engaging the attention of planners and policy makers all over the country. The diversity and importance of the economic and social roles played by tribal women has been exhorting the scientists and specialists. Indian expenditure on S&T has been meagre.

Suitable schemes of new technologies are highly inevitable for the upliftment of socio-economic conditions of tribal women. The basic objective of scientific policy resolution were *inter alia* attainment of technological competence and self-reliance, provision of maximum gainful and satisfying employment to all strata of society.

Project planning and project preparation are vital. Efforts to ensure this need, are to be sustained. Following obstacles are faced in course of their improving the standard of living through new plans and policies are : Improper training, insufficiency of technical staff and supervising facilities, lack of need based projects, non-availability of suitable mechanism to determine the appropriateness of the technology; insufficiency of government machinery to provide the basic needs like supplying the improved techniques according to the demand on time, non-involvement of local people and organisation, shortage of funds, need of proper S&T policy measures; non-availability of technical advise, rectifying and replacing the equipments in use, lack of political will. Therefore, accurate effort need to be implemented. The following measures have been undertaken for the success of policy and programmes :

1. Appropriate policy implementation from grass-root level.
2. Training facilities.
3. Repairing workshops.
4. Rewards for successful implementation of the policy.

Within the framework characterised by the difficulty of becoming a part of labour market, rural women have to face qualitative worsening because of great external problems affecting their personality. Now-a-days tribal women are looking for adequate working role enabling them not only to express but rather to find dignity through an action plan for socio-economic development. But the world of technology with its laws (more technology and less jobs) secures to be a further limitation to the fulfilment of the dreams.

Functional Literacy and Adult Education

Rationality and scientific temper does not subscribe to any absolute truth based on authority but world accept as valid only those things for which evidences have been collected scrupulously. It has no place for the simplistic (assumptions or superstitions) beliefs and prejudices. What is distinctive about the scientific methods is built into it, namely the possibility of disproof. It is an adaptive, integrative, negative feedback process where inputs are new data and new hypotheses as to what the underlying theories might be. Scientific knowledge does not postulate unconditional truths but truth hypotheses valid for a range of specified conditions. To act on strategies to be adapted to further the social relations and institutional perspective to poverty alleviation, and enhance capacity of NGOs in poverty alleviation on which organisation is most relevant to :

- Prioritisation of rights to resources, employment, just wage and basic needs.
- Adopt a multidisciplinary to perceiving development and problems and solutions.
- Building poverty and social relations sensitive poverty indicators and data base.
- Capacity building of NGOs to strengthen the capacity of poor people's organisations and withdrawal of direct work from their area of work.
- Handling the conflicts at community, family, local market level.
- Link support and advocacy work to grass-root action and networking.
- Strengthening expertise in capacity enhancement from social relation and institutional perspective.
- To combine expertise in political science, behaviour science and management studies to strengthen development issues, operational planning monitoring and evaluation, financial management for developing framework or methodologies for institutionalizing social relations perspectives.
- In or the organisational context the factors, goals and objectives, micro and macro strategies, leadership, human resource development structures, operational management, organisational learning, values and culture for appropriate selection of the area of operation, criteria and methodology, approach for achieving wider impact.
- Enhance sectoral strengthening efforts and enhancing accountability and learning of NGOs.

Above all to move towards such a level of empowerment, it is critical to acknowledge the linkage between the personal and political, between economic and political decision-making or resource allocation, use and control. It is important also on the basis of class, caste, gender and ethnicity as well as with state structure and policies. There is need to facilitate processes of advocacy for policy level changes in priorities and directions, to make them suitable to local alternatives emerging in different parts of the country in diverse context. Along with this, attention need to be given to institution building perhaps developing multiple organisations forms with their priorities and programme and access to information and sustainable technologies both in order to resolve these conflicts and help consolidate and institutional successes as sustainable equitable alternatives.

Lessons Learnt based on Past Experience

Local wisdom and knowledge cannot bypass nor be discarded if we want as sustainable livelihood :

- If women are given appropriate opportunities and enough room to bring out their strength, they exhibit tremendous power, against various exploiters. Their greatest weapons are initiative in taking action, perseverance and hope on one hand and strength to resolve conflicts as well as take on direct confrontation if the need arises on the other.
- Women's involvement changes the texture of the process of development by bringing out the needs of all groups in the communities. Traditional knowledge was based on building/augmenting/replenishing the natural resources as well as ensuring its use according to the livelihood priorities and on a value system that would be shared and used judiciously. As compared to that the modern system have promoted the value of the resources as commodities.
- Livelihood priorities such as drinking water, food, fuel, fodder, health services, etc., can be merged on a sustainable basis if their overall management is decentralised.
- It might be essential not only to change the schemes and programmes but also the institutional structures and roles of authorities that control these.
- More scientific inputs could really go into improving the traditional system and sources rather than totally discarding them and trying out experiments far away from the outreach of the community.

- Adequate quantity of water is given more weight age by the community rather than quality of water to begin with. Efforts on (health) quality aspects do follow quickly.
- Programmes for gainful occupation and development assistance in agriculture, industries and other economic activities.
- Development programmes which promote greater support by village men and, local institutions to women's lives. In order to make them more active and efficient in their participation in economic and social development particularly in the areas such as : Maternity protection, childcare facility, nutrition and health care, both preventive and educative, literacy and non-formal education, water supply and sanitation, simplification of house work, social and civic education, training, steps to increase efficiency and application of appropriate technology, health care, political will to improve. For ensuring greater participation of women in S&T programmes other important considerations should include : Building up of linkages with appropriate agencies, training of staff, implementation of experimental pilot projects in selected districts for effective propagation of appropriate technology, financial assistance for economic development and support to S&T programmes for rural and tribal women.
- Training should be imparted to develop and improves skill and competence of staff. The training contents, donation and requirements, of the trainee should be carefully worked out for making gainful employment in post training phase.
- Research is necessary input for successful implementation of any programme to be undertaken by rural/tribal women organisation. Exchange of ideas between functionaries of the schemes, the participants and those whose co-operation is necessary, through workshops, seminars, group meetings, group discussions, exchange of visits is necessary through workshops, seminar groups, demonstration on skill development and to provide a feedback to policy

formulation, and training components. There is need to formulate and implement a sizable number of centrally sponsored pilot programmes in one district in each state, union territory to develop integrated projects for tribal women with a multidisciplinary and multisectoral character. The main thrust on the structure of such a project would lie in providing an integrated service with minimum inputs in : (1) economic development in both agriculture and industry; (2) education programme particularly through adult education schemes and health service particularly better maternity care, and child health, nutrition, family planning, scientific culture and outlook can act as primary weapon to fight discrimination against women, women are the agent of change and men's equal transformation. Women scientists or entrepreneurs could develop or promote indigenous, appropriate technologies best suited to make the life of women less of a drudgery. The projects should create the necessary environments or institutional arrangements to take S&T to the doorsteps of rural families, in view of the absence of transport facilities.

The communication technologies should be well developed in operational management of environmental projects with participation, advocacy skills in the area of environment, knowledge of legal provisions, land resource management, bio-diversity, sustainable agriculture and promotion of linkages between groups and government banks, panchayats, research institutions, in the areas of environment and development and gender issues in operational management like knowledge of legal provisions and policies in the area of micro enterprises, promotion and strengthening ability of the groups to bargain with the state, markets (middleman traders, moneylenders, etc.) and to expand the control of poor over resources. The adequacy issues to provide technical skills like welding, carpentry, horticultural activities like mushrooms, poultry, floriculture and others. Introduction of new technologies to some extent has taken form of high yielding varieties, chemical fertilizers, and irrigation, access to information technology, education training, credit and resources, job and credit markets. In India of late there has been attempt to integrate the traditional and modern technologies for the effective development of economically weaker section in general and that of tribals in particular.

In order to unlock the human resources and to the main stream of national life, their present socio-economic status is to be improved, which has deteriorated during the last few decades under the impact of modernisation and industrialisation with special emphasis on adult education for women, through mass communication media, women should be made aware of their rights and privileges. Such awareness only will help them fight against exploitation.

Vision and Mission

New Thrust for

- Diversity pluralism and equal opportunity;
- strengthening relations with the equal environment work cooperatively with public and private institutions, legislative bodies, industry, state and community organisations and other units in the university;
- close linkages of faculty and students with the real world;
- interrelations between the various components of human ecosystem with the planning and principles.

Future Planning for

- Expansion of research and outreach with people, resources and facilities;
- postgraduate degree/diploma programmes for pre-service, in-service graduates;
- other training programmes — short-term and doorstep (tailored programme as per need).

Principle for Strategic Planning and Practice

- Improve access to quality education and expert knowledge;
- achieve more active learning;
- generate new knowledge and scholarship across mission;
- promote problem-solving to address for society needs;

- advance diversity within community;
- make people matter.

To conduct research towards improving the sustainable productivity and to support rural development sector the project on Women Agriculture and Rural Extension (WARE) will adhere with the NGOs through the collaborative research and training. The project will contribute to the institutional strengthening of voluntary private sector through the development and delivery of packages for playing significant part both as catalyst and prime mover in rural development. The main thrust will be directed towards youth employment, income generation and to generate mass awareness about development issues. The activities which are socially beneficial and economically viable like on-farm, off-farm, and non-farm will be taken up for the benefit of the community by doorstep and on station technology transfer through short-term training programme to ensure continuous benefits to the rural poor so that the resources and environment may be preserved for further generation.

In Specific Terms the Projects Aim to

- Impart relevant education and training to young men and women for managing income generating and development activities on behalf of rural producers;
- Offer training courses for policy makers, directors, general managers, and those in charge of specific managerial functions in rural enterprises and projects;
- Conduct research on operating problems in the rural sector in order to help improve the management of rural enterprises and projects; and
- Undertake basic research into the process of rural urban management to augment the existing body of knowledge on the subject;
- Technical programmers will include : Ph.D. in Human Ecology, M.Sc. in Human Ecology, Bio-diversity, Bio-technology and, Environment Management, M.Sc. in Women Agriculture and Rural Extension Management; Modification of B.Ed. and M.Ed. course.

Nodal Centre for Human Ecology

Financial Mechanism

For the projects, for the development of infrastructure for :

- Relevance of education;
- promotion of excellence and quality in education;
- equity and access for management of higher education and UGC administration for "Financial Mechanism".

Education develops manpower for different levels of economy and empowers the poor masses to become self-reliant enough to participate in the process of Nation development.

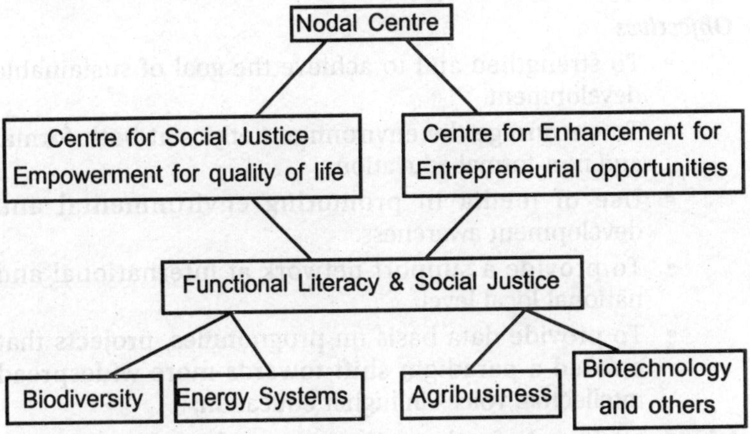

The Main Objectives of the Proposed Centre will be

- To identify various target groups of human ecology for women as well as those who are to be trained in imparting and in developing the field of human ecology and consequent, challenges/opportunities for teaching education in the field of human ecology.
- To review currently employed teaching methods and assess how far these are suitable, and what appropriate methods and training programmes need to be devised.
- To devise programmes and techniques for sensitising not only the institutions (the teaching faculty and students body) to imbibe human rights culture but also for developing sensitivity and respect for human rights at the community level.

- To identify activity and programmes in addition to running courses, that could be undertaken by the institutions for objectives of human ecology education.

The proposed centre will be located at the Institute of Home Economics, New Delhi.

International Centre for Human Ecology

Key Issues

To help people, help themselves through the practical application of knowledge and resources to improve their quality of life and that of future generations.

Objectives

- To strengthen and to achieve the goal of sustainable development.
- To encourage the environment ethics at both formal and non-formal education.
- Use of media in promoting environmental and development awareness.
- To provide a support network at international and national local level.
- To provide data basis on programmes, projects that created a paradigm shift towards more widespread intellectual voices in higher education;

taking certain tasks for them. The aspects of man's endeavours which was now called technology came about as a result of drawing together of the practical skills of craftsman and the method of seasoning development from a systematic investigation of nature, thus bringing together of practical skills and theoretical science created applied science. The technicists definition of technology explains solely in terms of its technical aspects while the humanist definition relates to societal needs.

MAKING THE PROJECTS WORK

Participation evokes the political idea that all individuals have a voice in making decision. Development is a charitable mission of industrial economics to alleviate economic and social hardships. Participating development suggests that rural villagers and urban dwellers with little or no expertise should have a say in introducing social service or in getting economy of their village or neighbourhood going.

Participatory evaluations do not pretend to meet the high standards of conventional evaluations, but they do not reflect the needs and responses of those whose life lives project must directly affect. They are less rigorous than conventional approaches. But they are more likely to provide information which can practically facilitate a project ability to meet its real goals.

Participating evaluations therefore offer a practical alternative to how project evaluations are conventionally done. This describes why and how : Why planners should consider this alternative approach for how to go about making this approach work to the best of advantage. This gives some background to the concept of involving beneficiaries in different phases of project managements. Participatory research has been key factor in project success. This describes how the methodology in participatory research diverges from conventional research, where they overlap, why they are different and what the implications are. The approach suggests how to put participants together into teams, how to train them in the basic of research, how to select methods that match their skills and concerns and how to use the results. This draws to show how concrete case of participatory evaluations have actually worked to contribute to project management to a wide range of professionals, consultants and beneficiaries. It describes an idea whose time has come. Development planners and managers will find a less costly and less complex way of generating useful information. Development evaluations will see the long-term benefits of turning users group into researchers. Project staff will find here an approach to evaluation that promises to improve rather than threaten a project operation, and beneficiaries will find a way of being participants rather than objectives in the evaluation process.

None could image how a poor farmer might apply the Horrod-Damor model or how a poor woman could pretend to tell the experts what was best for her children or her neighbourers or decline much to improve. Furthermore, part of the privileged wisdom suggested that most people in most villages and neighbourhoods ought really have little to do with building economics. They did not qualify as actors in the process of growth. Only those with resources qualified, and among them only those who could spend and produce in just the right way. This meant that villagers were doubly excluded. First of all it was really impossible for them to know the factors such things as the appropriate factors in the process of growth. Their action did not figure in the solution. It followed from this conventional wisdom that what these fledging economics needed most had little to do with what the majority of ordinary people thought they needed instead what they needed most according to conventional prescription, were high fielding privileged resources that in general only the most privileged people could put to best use. Developing physical capital took presidence over developing human capital. The idea appealed to development agencies since delivering physical resources was easier than developing human ones. Developing human capital was too risky. Creating village organisations through which villagers might have a voice in local development was too unpredictable a venture for more agencies whose officers had to struggle against expenditure dead lines.

If things have changed, and they have in some measure, it is not only because the deadlines have changed or the ventures proposed by participating developers are less risky for donor agencies. It is because they work better than the conventional approaches. A village organisation which initiates the income generation scheme itself tries diligently to keep it going. They work better because when beneficiaries have state in an investment, the investment usually pay off.

Background to Participating Development

With the effective initiatives of FAO an important turning point is in the acceptability of engaging beneficiaries in the design of development assistance. The outcome was to encourage FAO to further act in poor project design and maintenance under an umbrella organisation as the People's Participatory Programmes (PPP). The PPP began to design projects with beneficiary participation in mind. Now ILO actively advocates the implementation of programmes on Participatory Organisation of Rural Poor (PORP). The United Nations

Development Programmes (UNDP) : Partners in development programmes has made annual grants to community based organisation since 1988 to support local NGOs which in turn design and implement participating projects. Nearly half of the projects financial by UNDP for women (UNIFEM) administered through NGOs are monitored to ensure beneficiary involvement. In short to evaluate and one of the central ways participation functions in projects is by providing community organisations tools to carry out then own evaluations.

Participatory projects are more common now-a-days, than even before, arising frequently out of circumstance where more conventional approaches have failed. They have provided poverty relief to the poorest communities. Where traditional development strategies have failed community participation and project evaluations are successful.

In the course of assuming more project management, community groups and individuals gather and apply information, learnt from experiences. They learn from experience. They begin to trust their own knowledge. Past experience on the income generation in mushroom and floriculture have revealed that when beneficiaries become evaluators the results are after more revealing, information more reliable, more benefits accrue to the projects and project designs are more appropriate and enduring. Looking to beneficiaries, peasant women and men and other horticultural activities and small merchants for expertise challenges what most people know about development as well as the way they normally come to know it. It turns knowledge upside down. Knowledge is no longer the ethereal substance distilled from books and lengthy surveys but is rather a useful body of information coming from most obvious places.

The past experience based on action research project reveals that one of the principal functions of the organisations through which participation in project implementation is to acquire information and use it in other words to evaluate the community participation in project evaluation and implementation tool to carry out their own evaluation. Along with the body of knowledge and research it has generated three threads of thoughts. Each has a special impact on the emergence of participatory research. The first thread on the specific problem focused projects were found quite practical and the information was quite accessible and could be used by the community. The second thread of objectivity of social transformation could be achieved, which was necessary step towards participatory action research, participatory rural approaches. The second string of collective learning as a political act capable of rallying to take action against injustice. The third thread the disenfranchised step a theoretical one between conducting participatory evaluation and the practice of conventional social science which is neither very practical nor explores in depth the epistemological challenge which participatory research poses to social sciences. This shows how participatory research forms the 'new' physics in challenging the fundamental assumptions in social science in challenging especially the rigid destination between subjectivity and objectivity and analytical pre-eminence self-interest enjoys in research on human behaviour.

It has also been observed that poor communication affects the technical as well as the social aspects of the project. The innovative evaluation strategy could help project personal and beneficiaries, to communicate and train local individuals in basics of social inquiry and have them live in midst of project beneficiaries and staff. It worked so well in fact that to use this approach in other areas and other projects, especially in communicating the marketing strategy. The main funding of the project especially mushroom technology, its marketing strategy and consumption pattern, with the local other actors of mushroom industry were found quite useful and middle man had convinced mostly the growers that the price for mushrooms was higher than the price given by the marketing cooperative. The only one to form the cooperation were too few members to operate effectively. The measuring impact was the secondary concern. The primary was to provide a mechanism by which information could flow from the beneficiary to management. With some exceptions, it is a useful means to make protect work better. Evaluation does not mean performance review although implicitly, a performance review occurs, instead, of a collective examination of problems aimed at solving them.

Project Evaluation Especially

Donor focus evaluations assumes a clear arrangement of these roles in a ladder of control which plays the beneficiaries at the bottom. They reinforce authority of wealth of donor and the science of evaluator. In beneficiary focus evaluations such roles are never given in advance. Their ultimate objective is to create a solution in which beneficiaries not donors can act with authority.

The Project Manager's Dilemma

Introducing innovations into cultural and economic environments which are quite different from one's own is a complex task and things are bound to go wrong. On the one hand they must plan these endeavours in detail while on the other hand they must be prepared to be flexible to learn from mishaps and to change course midstream. While on the other, manpower must impose rigorous schedules and cost conscious management. They must practice an administration which fulfils at one and the same time irreconcilable objectives. They must be both highly flexible and strictly accountable. Of all these beneficiary involvement in evaluation activity seems most likely to lead towards a management style which builds the capacity of project to learn. Participation in other components has its rewards; where beneficiaries share costs project cost less and where beneficiaries give labour, project maintain some contact with labour, engage in evaluation research, the rewards are greater and more lasting and the participatory ideal is morefully realised.

'Open ended' and 'framework' approach : An open ended approach which interferes minimally in the questions which group members choose to include or a *framework approach* which stipulates areas of enquiry that ought to be covered. An open ended approach encourages more autonomy and commits a group more to grappling with question they choose. They learn more about their ability to learn then they do about managing their courses better. On the other hand a framework approach excludes wayward questions it builds the experience of a decade of participatory research in one state of project. Participatory evaluations fill a very practical need. All development assistance project that have anything to do with people must adapt to changing conditions. Beneficiary needs change. Dilemmas arise which projects have to confront and solve. No beneficiary may emerge, when this happens project managers need to get information and then make decisions to solve a crisis or steer projects in different direction and none of this is easy. In a single exercise, project participants acquire knowledge and become the agents of change, this is the first step in the process of achieving sustainable development.

Participatory Evaluation — A Practical Alternative

Participatory evaluation fills a very practical need. It offers a practical alternative to how project evaluations are conventionally done. Why and how planners should consider this alternative approach and how to go about making this approach work to the best advantage. The idea is relatively new even controversial and it requires some justification. Participating evaluation relies on recipients, project staff, and beneficiaries as researchers instead of professional evaluators. This changes evaluations in critical ways. It recognises recipients, project staff and beneficiaries as key custodians of knowledge about what projects do and do not do. These individuals are a project's most concerned stakeholders. One of these most concerned stake-holders are engaged in project successes and failures, they have taken the first step forward assuming some responsibility for implementing changes. This is an important step towards attaining a sustainable development. Participatory evaluations do not pretend to meet the high procedural standards of conventional evaluations, but they do reflect the needs and responses of those whose projects most directly affect. They are less rigorous than conventional approaches. But they are often more likely to provide information which can practically facilitate a project's ability to meets its real goals.

Participatory evaluations therefore, offer a practical alternative to how project evaluations are conventionally done. It is more functional practical advice and suggests how to put participants together into team, how to train them in the basics of research, how to select methods, that match their skills and concerns and how to use the results to show how concrete cases of participatory evaluations have actually worked to contribute to project management.

Development planners, managers, will find the way less costly and less complex way of generating useful information. It is an approach to evaluate that promises to improve rather than threatens a projects operation and the project find a way of being participants rather than objects in evaluation process.

Institutionalisation

Government of India has been drawing much attention to science and technology programes ever since its independence. The problems of integrating tribal women in the development process are now engaging the attention of planners and policy makers all over the country. The diversity and importance of economic and social role played by tribal women has been exhorting the scientists and specialists.

Suitable schemes of new technologies are highly inevitable for the upliftment of socio-economic conditions of tribal women. The basic objective of scientific policy revolution were *inter alia* attainment of technological competence and satisfying employment to all strata of society. Project planning and project preparation are vital efforts to ensure this need.

- Translation of these voices through educational programmes in human ecology to be viable, dynamic in creating, imparting knowledge through teaching, research, service and outreach.
- For the formulation of national education policy with respect to human ecology to strive for its implementation to achieve its goal.
- Education and development programmes to involve women right from grass-root level.
- To coordinate network of regional and national centres and to provide a forum for gathering and dissemination of information on experiences and findings of those in the field.
- To facilitate the entry of women into employment in the field of science and technology and their progress within such employment.
- To deal effectively with global development issues and opportunities to broaden their perspectives.
- Training and award programmes to support activities to

meet both needs ultimately leading to the production of critical mass of experienced trained researchers for global, sustainable equitable development.

SETTING PRIORITIES FOR ASIAN CENTRE

To act as a catalyst for resources needed to create financial and administrative infrastructure needed to undertake long-term research, agenda to ensure appropriate research priorities :

- International collaborations for building scientific, academic and development communities by linking its national partners and networking with international collaborators to increase and contribute to the international research.

Mandate

To initiate, encourage support and conduct research into the problems of the developing states of the country.

Mission

Empowerment through knowledge.

The proposed centre will be attached to the Nodal Centre.

University Centre for Human Ecology

Introduction

To facilitate the development of capabilities of the community for the overall development of the national and help people to improve the quality of their lives. Social development of people needs organic linkages between education and society. With the view of present education system and its human resource development activities programmes for the total population related to the socio-economic development. Delhi University has forwarded several projects to facilitate linking of higher education with the development process, confirming to the UGC policy of considering extension as the third dimension of higher education which are need based flexible and catering to a large number of people with diverse needs.

The project proposal entitled "Studies on women agriculture rural extension through education and training programmes with special reference to pre-service graduates" forwarded by University of Delhi to UGC to establish Centre for Adult Continuing Education, Extension Work and Field Outreach for the optimum involvement of teachers and students where departments, like Home Science, Microbiology, Biotechnology, Physics, Sociology, Psychology, Economics and school of social sciences will be the integral part of the academic work. The basic aim of the project is to adopt an integrated approach to help sustainable linkage with all sectors of community, consisting all age groups with the institutional resources to the community and to gain insight from contacts between knowledge resources and socio-cultural realities with the view to reflecting these in the entire curricular system of higher education including outreach and research. It is a two way process between the experts and the people and intellectual intervention in the community living problems which need to be overcome through the educational process.

Aims and Objectives

Impart relevant education and training to young men and women for managing income generating and development activities.

- Offer training courses for policy makers, directors, general managers and those in charge of specific managerial functions in rural enterprises and projects.
- Conduct research on operating problem order to help improve the management of enterprises and projects.
- Undertake basic research into the process of rural urban management to augment the existing body of knowledge on the subject.

After finishing the course the participants will be able to :

- Relate extension education to the development issues.
- Design curricula and programmes that contribute to development.
- Administer curricula and programmes, selecting for different target groups appropriate methodologies and media.

Rationale : Reaching the Unreached through Education to Improve the Quality of Life

Programme/Activities proposed :

- Continuing education courses for the university group.
- Adult Extension Education for those eligible for university based courses.

Research

Role and functions of the Department of Adult continuing extension, education and field outreach.

Catalyst Role

- Acting as focal agency in the University system.
- Serving as technical resource centre for universities community activities.
- Assessing the need of the community.
- Developing curriculum and multidisciplinary University programmes, involving faculty students from various disciplines.
- Stimulating other departments of the University to undertake similar activities with respect to their discipline for continuing education of graduates, extension education and field outreach.

Teaching, Training and Research

A project proposal entitled "Research and Training in on-farm off-farm and non-farm sectors for human resources development with special reference to tribal women".

- Teaching programmes/courses in adult and continuing education.
- Conducting need based continuing education courses not undertaken by other departments of the university.
- Conducting activities/programmes for upgrading knowledge skills or acquiring new skills — short-term diploma certificates programmes.
- Training short-term aimed at various groups which are skill income generation, entrepreneurship development, family life education. Para legal training for grass-root workers/activities.
- Community outreach activities — responsibility to reach out to society, specific disadvantaged groups, *e.g.*, bonded labour, child workers, urban slums and villages.

Programmes

Vocational/Career Oriented Programmes

Empowering people at different socio-economic and educational levels with marketable vocational skills, which are appropriate to the market demands both in rural and urban areas. For students who have undergone general education courses in continuing education could offer specific skills. For those already in jobs to update skills or acquire qualification for career advancement.

Equivalency programme. Catering to parallel programmes for disseminating knowledge in all walks of life and for different segments of the population through need based and relevant education programmes to fill in the gaps in their intellectual growth, enhance professional and technical competence and promote undertaking of contemporary issues.

Quality of life improvement. Upgrading living standards and lives of citizens to an accepted level of development and focussing mainly on the disadvantaged sector for this purpose.

Developing the Discipline through Academic Programmes and courses.

- Foundation course.
- Course integration with different disciplines of conventional academic university departments.
- Projects related to extension field outreach in-built into a discipline or studies in conventional academic departments.
- Certificate/Diploma courses for adult education personnel in different spheres.
- Undergraduate and postgraduate level optional courses in adult and continuing education.
- M.Phil./Ph.D.

Academic Credit

The work of students involved in adult education, continuing education, extension and field outreach work, should be considered for academic credit. For determining academic credit :

- Project related to extension/field outreach within a discipline at undergraduate or postgraduate level.
- Field work in respect of a subject of study.
- Participation in literacy, post literacy, continuing education under the "ground level model".
- Evaluation as per university norms.

Collaborating and Networking

- Networking with government departments and other agencies of the country.
- International collaborations — to implement the programmes for higher education and to draw the programmes of collaboration by the participating institute — for linking higher education in the priority areas of environment, population education and science — and technology and gender issues.

Documentation Particularly of Field Experience to Enrich Theory and Dissemination of Information

- Preparing material.
- Disseminating information.
- Use of computers and internet — Publications.

Monitoring and Evaluation

- Monitoring and evaluating the Department's activities as well as ongoing extension in the university.
- Reporting to Nodal University.

Proposed venue of Project : South Campus, University of Delhi, Delhi.

Projects

Project No. 1

Studies on women agriculture rural extension through education and training programmes with special reference to pre-service graduates.

Description of the Project

The project will contribute to the development and delivery of 3 courses and Ph.D. programme (1) M.Sc. 2 years, (2) Pre-service (9 months), (3) In-service (3 months) courses on rural extension. The course material will be developed, experts both foreign and local will participate in the development and teaching, fellowships will be granted to the participants from India and in a later phase from Pakistan, Bangladesh, Sri Lanka, Nepal and Bhutan.

Most modules have a multidisciplinary approach, integrating the different agricultural fields, special interests of the participants will be attended through the special project module for study and research activity.

In total a number of 500 extension workers of which atleast 75 per cent women will be trained in order to ameliorate the position of poor rural women in the respective countries.

Project No. 2

Project Title : "Transfer of agro-biotechnology with special reference to horticulture for sustainable integrated organic farming's system."

Project Summary

Rural Indian people depend on agriculture as their main occupation. Therefore, it becomes important to generate and transfer such technologies which will benefit agriculture. Women work as agricultural labourers in the fields as well as in their homes. Analysis has suggested that the caloric intake of landless and marginal farmers is far below the recommended dietary allowance.

In view of this it is essential for rural men and women to go for additional income using simple technologies utilising the limited and available resources for maximum production.

Individual technologies are sufficient to make a significant impact by judiciously integrating various technological inputs to suit these sources available in terms of land, capital and human resources.

The necessary synergy can be achieved in bringing visible and sustainable benefit to the target groups. Several site specific technology packages can be worked out under this project, like agro-biotechnology packages for very small plots either/around the households or in small farms have been worked out with focus on floricultural crops.

Floriculture is the major income generating source even on relatively small plots. However, the other needs such as food, fodder, fuel needs of the household can be fulfilled by appropriately integrating these both in terms of space and time with floriculture. Further all practices, *e.g.*, the selected fruits and vegetables and pulse crops would, enhance the nutritional status. The residues would be recycled for feed and fuel purposes. The necessary manure will also be produced locally by composting.

Following activities shall be undertaken under project :

1. Cultivation of pulses for research and development rotation, green-manuring and inter cropping, post harvest technology, raising sprouts, recycling of agro wastes for mushroom cultivation and vermi composting
   - Experiments to be conducted at the demonstration site.

(i) Cultivation of Moong/Arhar for post harvest processing, for sprouts raising :

(a) Rhizobium — to inoculate rhizobial species for root nodules of pigeon pea plant.

(b) Acetobacter and azetobacter : to fix transgenic azetobacter which can fix more nitorgen than the normal strains.

: to increase 3 fold synthesis of ammonia NH^+

Mycorrhiza : Total phosphorus organic carbon to have dense association of VAM fungi to inoculate in moong and moth.

Blue green algae — application of different cynobacterial innocuoulants alone or in combination with fertilizer increase the yield.

(ii) Recycling for manures : vermicomposting green manuring crop culture of organic farming.

(a) Transfer of technology for the production of Bio-fertilisers.

(b) Mycorrhiza — Technology suitable to almost all crops including plant.

(c) VAM — In the farmer's field to increase total phosphorus, potassium content and pH of soil of crop fields and experimental plots, to have dense association of VAM. Feasibility studies on the production, product development, recycling to help grass-root entrepreneurship.

(d) Blue-green algae — In the farmer's field to increase the nitrogen content of soil.

(iii) Utilisation of agriculture waste — Mushroom production liquid and solid compost production.

Mushroom cultivation. It is beneficial by all means for rural people. Women who work as agricultural labour can also grow mushroom. It does not require much of attention throughout the day. Cultivating mushroom is an ideal income generating activity for small scale and landless farmers which besides generating income also supplements nutrition. Mushroom is a horticultural crop grown worldwide for income generating on cottage scale as well as for commerce. It is also a perfect health food recommended for use to enrich diet with vegetable, protein, vitamins, minerals and fibres. With the right training women can go into this venture and supplement their family income who contribute a high percentage of the labour in the field, especially in the developing countries.

(iv) Horticulture — Integrated farming system and intensive land management :

(a) Crops — fruit-ber, papaya, kachnar.

(b) Vegetables crops — beans, cluster bean, French bean and cow pea.

(c) Exotic vegetables — cucumber, chilies, mint and methi.

(d) Flowering crops — bulb crops — gladiolus and tuberose.

(e) Seasonal and annual and perennials — Gypsophilla salvia, lupine, larkspur, tagetes, motia and others.

(f) Seed and planting stock will be obtained from near by agricultural universities research stations, village-based nursery and fruit utilisation department.

(g) Training the women in floriculture.

(h) Integrated nutrient management in soil.

(i) Recycling of Agro waste — By several eco-friendly composting methods, Biomethanation processes, and water purification for human consumption.

The programme facilitates the poor community to increase their skills, income and create additional job avenues for self-employment and help in improving their health status.

Project No. 3

Conservation of Ecosystem of Central Himalayas with Special Reference to Gola Catchment Area.

Project Objectives

This project supports the main objectives of the Government of India's strategy for the development of its forestry sector into popular disciplinary. The objectives are to increase afforestation programmes, preserve the ecosystem and strengthen forestry research, education and training. The project links socio-economic needs of the villagers (through appropriate enterprise promotion and conservation training) that aim to protect, enhance and conserve biodiversity. The project is also consistent with Indian National Environmental Action Plan and National Bio-diversity Action Plan.

Location of the project is yet to be decided. However, the linkages will be established with International Center for Human Ecology.

Project No. 4

Training in Off-farm and Non-farm Sectors for Human Resource Development with Special Reference to Tribal Women.

Objectives

Institutionalisation on sustainable basis — The training programme on-farm, off-farm and non-farm sectors for human resource development as a whole.

Linking the curriculum from grass-roots to university level for commercialisation and management.

Purpose

To assist University Departments/Research Institute Professional Bodies/ Individual Scholars for conducting : (i) Research including action research; (ii) Workshop/Seminars and; (iii) Publication of Studies and documentation.

Science and Technology can be utilised to upgrade traditional skills, introduce new opportunities of work, reduce drudgeries faced during various household chores, improve the environmental and living conditions, reduce health problem. In fact, Science and Technology is capable enough of solving many problems of the people, if applied amidst their own environmental and cultural contact without creating any social and economic imbalances.

Environmental protection and development are the two facets of the same major issue. Environment and development are interdependent and cannot be separated. The questions related to environment are closely linked to human economic and social activities. The development of mankind cannot be sustained if neglected. Likewise, protection of environment in isolation from economic development and technological progress has no basis. Therefore, we must trace back to the root cause of environmental degradation and find the best way to protect environment integrating it into economic and social development.

- To assist those reputed social science research agencies which have a proven expertise and willing to undertake purposeful studies on the economic development, problems and requirements of Scheduled Castes and also on the impact of the implementation of various economic programmes undertaken by the governmental bodies through instruments of economic development.
- To assist and develop linkages with governmental organisation for undertaking research including evaluation studies/projects, workshops and seminars helpful in orienting developmental programmes and disseminate knowledge and exchange of experience and publication of literature on tribal development.

Entry Point

Approach for environmental education, training of school level teachers, University staff, trainers etc.

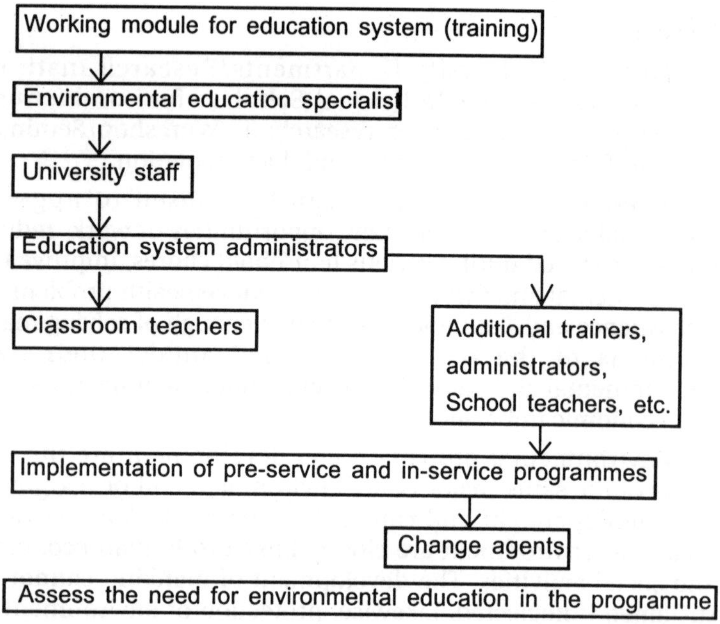

Extension

Linking Extension, Education and Research

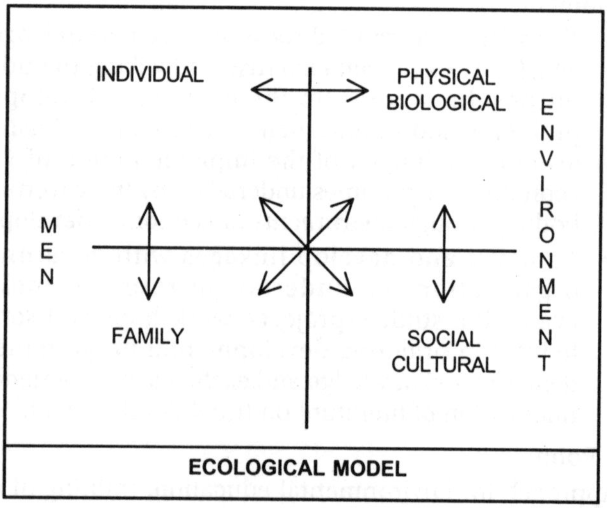

Ecological system model in totality focuses on family ecology. Center for Human Ecology will function ecologically emphasising collective efforts to achieve a critical mass of knowledge and activity, energising the field and contributing to a better quality of life of all. It will be the combination of scholarly productivity and modelling.

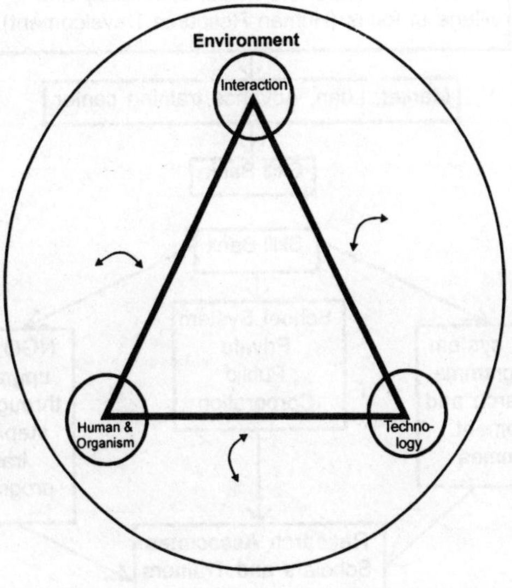

Farm household system in Indian rural setting consists of the following sub-systems which are closely interlinked and inter-active.

- household as decision-making unit;
- in-farm, off-farm and non-farm activities.

Flow Diagram

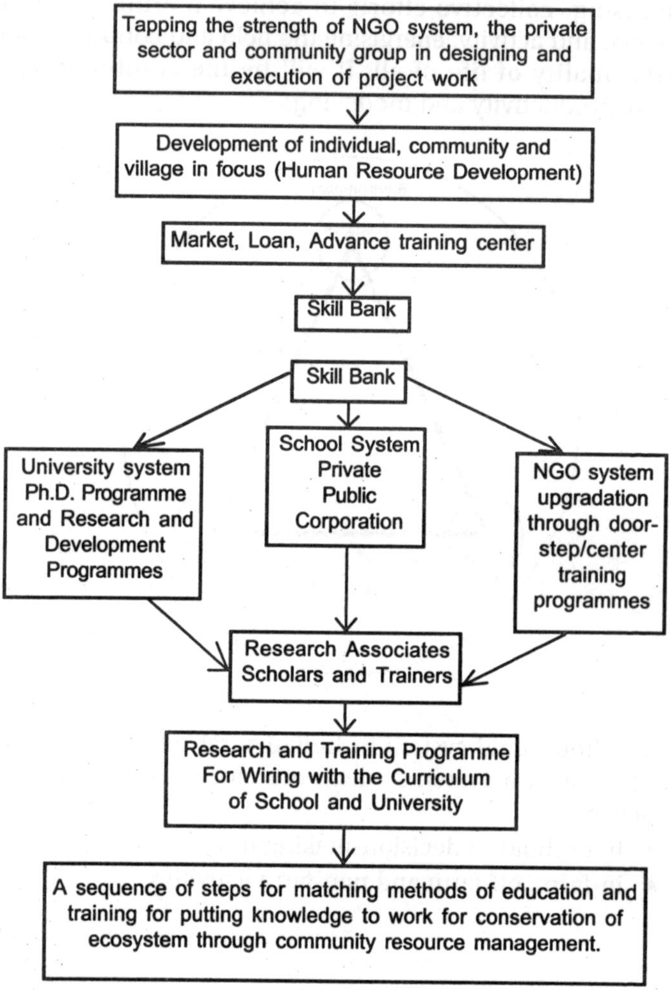

Methodology

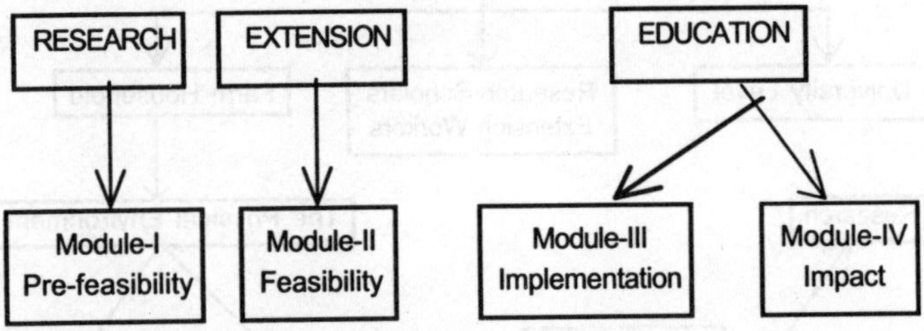

Environment
Climate
Soils
Topography
Water
Vegetation
Infrastructure

Policy/Institutional
Environment
Environment
Research, extension
Support services
Socio-cultural
Environment
Community Culture
and Values Tradition

Target population
Identification of
Ecological activities/
farming system
Resource mobilisation
and provision
Management counsultancy
Linkage building with
third party
Process extension
and movement building
self-evaluation Marketing

Participatory

Multi level

Economy supporting
the poor

Rate of adoption

Income generation

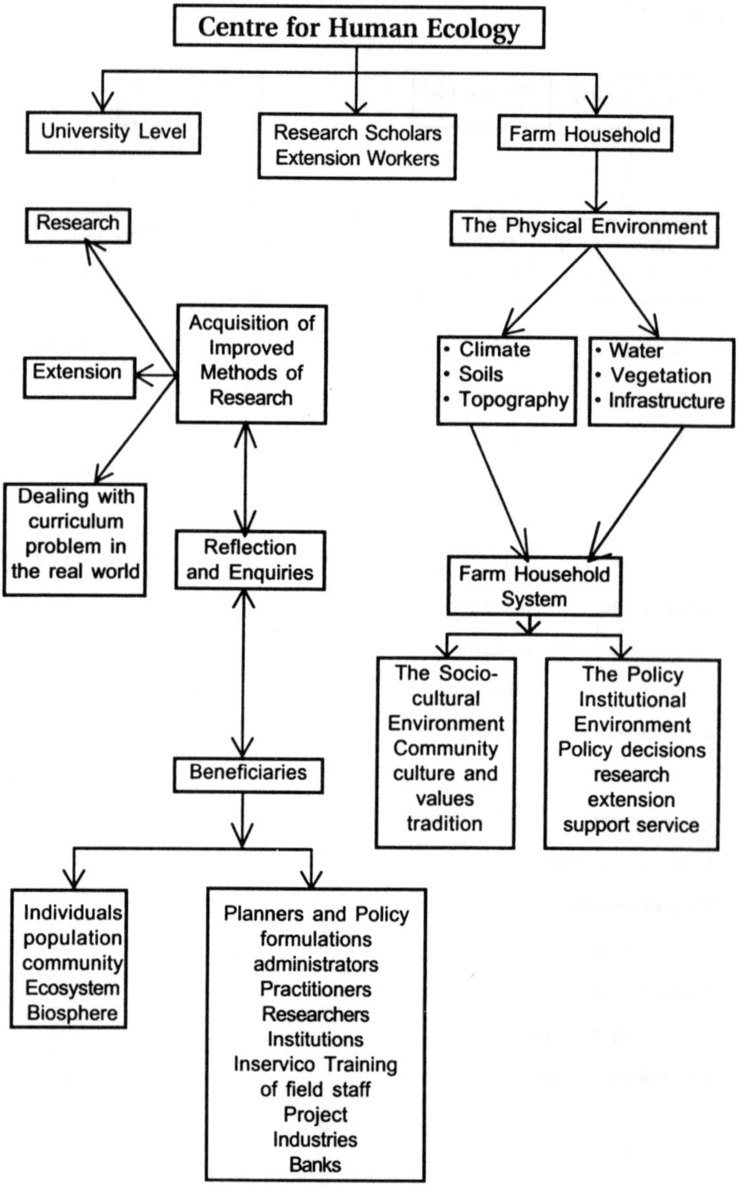

University level approach for reaching the unreached through research, extension and education.

Making Entrepreneurs

A society needs higher achievers and economy needs entrepreneurs for development, whose contributions through their manufacturing enterprises have created an impact in the market, in product, in technology and in the growth of this sector. They are impact makers, recognised by name of the products or their enterprises indeed it is the enterprise which ultimately creates employment, generates income and contributes to the economic growth. But it is the entrepreneur behind the enterprise, the individuals, the human, who is responsible for conceiving the project, formulating and implementing it successfully, running it in a manner, and style of his/her own facing uncertainties and taking risks, withstanding the stresses and strains.

In the Indian growth scenario we have come a long way in the last couple of decades in widening the base of entrepreneurship through our policy commitments to indigenisation, development strategies for opening up industrial opportunities and creation of an elaborate support system to remove constraints on new entrepreneurs. It is essential the emergence of the new entrepreneurial class and their achievements and contributions to the industrial growth are documented and analysed through research and society at large recognises them key contributors, to development and respects them as a special breed of individuals, possessing personal qualities and skills which not even one possess and without whom economic activity in the form of enterprises, industrial or otherwise, cannot take place and therefore, production, income and employment benefits cannot materialise.

In short country needs many more entrepreneurs, as many as can emerge from all walks of life, from all regions, to exploit existing opportunities and to create new areas through innovations, imaginations and creative urges. If entrepreneurs are required to set up enterprises, success and failure of enterprises is a result of the capabilities and competencies of these entrepreneurers. The growing menace of enterprise failure has taught us several lessons, mere promotion of entrepreneurs and creation of enterprises is not sufficient. These enterprises must run profitably and grow if the economy is to drive the benefits from investment of scares resources in them. Experiences of financial institutions and assistance agencies now reveals that a major cause of enterprise failure has been inadequate capabilities of the entrepreneur to manage the enterprise. Quality of entrepreneurship, and therefore, the person behind the projects are acquiring, greater significance than the viability of the opportunities. Research has, therefore, to catch up with the growing needs of improving our perception and knowledge about what makes successful entrepreneurers. This knowledge resource is going to be all the more important in the 21st century as the Indian economy is becoming increasingly competitive more open, less protective, more dynamic in terms of technology and change and less secured for all those owners of business who could survive despite entrepreneurial lacunae, because assured markets and high profitability protected them from early exists.

For years many entrepreneurs and business man individually and through their associations have always complained about rather "hostile" environment in which they have to enter and interact critical scarcities fluctuating policies, initiating procedures, and formalities, time consuming and costly delays, bureaucratic attitudes and industrial administrations have been repeatedly cited as constraints on entrepreneurial and business entrepreneurs have grown rather rapidly and have made an impact in the market. Who are they? How they achieved notable successes while most have not? What kind of personal and family.

Entrepreneur Opportunities for Women

More than six hundred Organisations/institutions in India are directly and indirectly engaged in propagating, developing, promoting and conducting entrepreneurship development programmes, training, however, has been the most important component of all such entrepreneurship development programmes. Such trainings are most commonly focused on four major areas — entrepreneurial quality/motivation development, enterprise launching including resourcing; enterprise management, social responsibility and entrepreneurial discipline. Generally trainers for each of these areas require different orientation and specific skill related to entrepreneurship particularly for entrepreneurial quality/ motivation development a capsule course under the title "Entrepreneurial motivation training".

Entrepreneurial motivation development includes proportions, objectives, psychological conditions, conducting and evaluation of EMT and training methods — concepts and application, different methods applicable in knowledge, skill and attitude. In total dynamics of motivation includes; understanding motivation process. Entrepreneurial motivation, achievement, motivation, significance of achievement motivation, characteristics of person with high need of achievements and achievement motivation *vs* entreprenurial motivation.

In the development process two major factors have played significant role in developing entrepreneurship one of them is development of human factor, the entrepreneurship himself, another major factor is development of environment where entrepreneurial activities can flourish and grow.

The human factors, refer to the attitude, desire, and motivation of an individual, his capacity to perceive, the environment changes and opportunities as well as his ability to solve.

Entrepreneurship Development in Agriculture

Dynamics of entrepreneurship is the character practice and/ or skill of an entrepreneurer. A entrepreneurer is a person who organises and manages and assumes the risk of business. Accordingly, entrepreneurship referees to identifying/innovating ideas/products and services; mobilising resources, organising production/services and finally, marketing them covering the risks with constant strive for growth and excellence.

Stage of Growth

In the context of employment generation, the terms, *i.e.*, entrepreneurship, self-employment and income generation are often used interchangeable. Although there a lot of commonalties among these concepts, yet the three terms are not the same. Self-employment refers to full time, involvement in one's occupation or pursuits in which one may or may not have to take any risk to mobilise imputes and other resources to organise total production and services or to market the product and services.

All entrepreneurs are self-employment and income generating persons. All self-employed and income generating persons may not be necessarily be self-employed. The three, however, can be viewed an initial, middle and terminal stages in entrepreneurship growth process.

The 1st generation of entrepreneurers requires to develop :

1. Entrepreneurial quality/motivation.
2. Capability for enterprise, launching/resourcing.
3. Ability to enterprise management; and
4. Sense of responsibility.

Entrepreneurship development strategy follows a cycle consisting of stimulatory, support, sustaining activities. The growth and process can be further accelerated through planned effort, by integration of stimulatory, supportive, and sustaining activities. Training has been accepted and found very effective intervention in motivating and developing entrepreneurial qualities, capabilities, and ability for enterprise launching and management, irrespective of educational background and upbringing.

Micro-enterprise for Women

Entrepreneurship and business management theory have been highly developed into conceptual frameworks by academic institutions. There is need to tie this into the reality of managing micro-enterprises.

Income generating activities are a major point for NGOs concerned with women's development. It is recognised that women's economic development is a powerful precursor to women's empowerment. NGOs and support agencies need to develop skills in identifying viable activities based on market opportunities availability of raw materials, relating them to local conditions of women's knowledge skills, attitudes and need for training. The conceptual issues cover :

- The micro-enterprise covers small units, either in manufacturing and service sectors, usually managed by one person; profit orientation for survival and growth in the context of poverty alleviation, in both rural and urban areas.

- For creation enterprise social inequality can be reduced by empowering women and economic independence is a step in that direction. Women need training in confidence building, motivation, assertiveness, and negotiation skills for generating economic opportunities. The inputs are also required in financial, technical and managerial terms to set up their own enterprises. Both collective and individual approaches are required in training, counselling, to suit the needs and demands in a given situation. For greater impact community development attempts to improve the economic status for women should be taken up in synergy with health, housing and education. Government support agencies need to upgrade their communication skills to disseminate features of available programmes appropriately and explore new opportunities.

Enterprise management requires counselling services at various stages, before and after start up of their enterprises. NGOs, development skills, to include functional areas such as finance, marketing, production and managerial skills, such as negotiation, time and quality management and logical approach to problem-solving.

Support from NGOs and other agencies should not make women dependent on them, rather such support should be reduced in stages.

For entrepreneurship training academics and practitioners should create a knowledge base through interactive research, to help identify needs of women entrepreneurs, and with the help of groups including support agencies, to provide mechanism for building strengths for survival and growth. Government, NGOs support agencies, and intermediary organisations need training to be more entrepreneurial in delivering their services, and need training opportunities to develop communication, motivation and counselling skills. Training programmes should be need-based with objectives in 'need to know' modules and should not be targeted only towards numbers trained. To measure impact and success it is essential to follow up training programmes and provide more guidance and information to encourage start up of enterprises.

Public policy issues need to be modified urgently to make it easier for women to work at field/decision-making/policy levels. Officials of banks and other credit agencies need training in order to become sensitive in their attitude for fostering growth of women's micro-enterprises.

Alliances and links between large, small and tiny sectors should be encouraged with input on quality, resources and marketing channels. Limited resources will deliver better value for money if integrated development programmes and income generation on a sustainable basis. The women's movement will benefit by enlisting the support of men at all levels, and promoting a gender partnership to improve the quality of life. The educational system from an early stage should encourage attitudinal and cultural change in the society towards women's status, equality, empowerment and entrepreneurship development in general.

Micro-enterprises

Importance of small and medium scale enterprises for economic growth, industrial development and employment generation at a relatively less capital cost is now well recognised in developing countries. In a small enterprise, the central figure is "entrepreneurer" the person who catalyses researches like risks and manages them so as to establish a viable, sustainable employment generating entity. He/she, therefore, need to be identified, motivated, strengthened and supported. A dearth of such entrepreneurs is a critical problem in many countries. For various reasons, entrepreneurs do not spontaneously respond to available business opportunities despite various inducement, schemes and programme to promote and assist them. An effective mechanism is, therefore, sought after to enlarge the pool of local entrepreneurs. In countries where economic activities have been historically dominated by few facilities, caste, community or where educated manpower is not fully utilised, therefore, are problems as well as opportunities for developing a new breed of entrepreneurers from non-business communities and tribes, the first generation entrepreneurs, who can become agent of growth and change.

International development agencies providing assistance to developing countries have often found "a lack of mechanism" to ensure that the people who received funds, training, services are those who will actively utilise and supply them. National/State level development banks are similarly faced with the problems of a large number of business failure or loan defaults, and one of the key cause is located in the lack of entrepreneurial/managerial capabilities in the promoter/owner entrepreneur.

The World Bank has been attracted to new initiatives in small enterprise promotion, particularly in developing countries.

"Having recognised the importance of small enterprises as essential for activating latent entrepreneurial talents" in less developed regions and to initiate "a widely diffused yet viable process of industrialisation in less developed countries". The nature and characteristics of institutional structure required for small entreprise promotion and development has been one important area of concern of the bank. Credit is realised, is only one facet of the problem. The vital facts are identification of viable small business opportunities, provision of required managerial and technical assistance and supply of competent local entrepreneurs to take advantage of business opportunities and set up and run viable industrial or business ventures.

Some of the Indian experiments in the last few years in the small industries development strategy have been well documented in the focus of the achievements of entrepreneurship development programmes in diverse conditions in India and experiences of the several developing countries suggest that potential entrepreneurship in setting up and running viable small industries is lying untapped in a variety of communities, caste, and occupational group and this can be effectively brought out and developed by imaginatively conceived and effectively carried out entrepreneurship programmes (EDP). A comprehensive training-cum-development programme can seek out and convert non-industrial and industrial employees, technical and non-technical graduates, small traders, and unemployeds, artisans, women and sons of farmers, or landless at low cost; as fairly well demonstrated by experiences in various parts of India and some developing countries.

Economic Development Alternatives

There is ethnic mix of population in the country. The overall economic background of the people, the availability of infrastructure facilities, technology, managerial skills, and the market potential, there are also some limitations to development due to geographical, topographical, and resource constraints. Another aspect is the availability of proper human resources with necessary skills. There is also the strong feeling that any development should not be at the expense of disturbing the ecological balance of tribal/community life or at the cost of plundering of forest resources, considering all these aspects, the following issues may be worth considering :

— Ethnic village community/tribals still following their old traditional ways of living and not very much affected by the modern civilisation.

— The old communities/tribals are gradually moving towards modern civilisation.

— Refugees, mostly hailing from other states and settled in metro cities/islands.

— Inhabitants who came from various states of mainland for trade and business.

— Persons engaged in gradually developing the tourist industry because of the diversity of various groups settled in any common strategy of development of entire population will not hold good because of their varied background. At the initial stages of development, where appropriate skills are lacking, there will always be a tendency to import human power from outside. There will have to be different types of training and education institutions for different groups based on their level of current education and training.

Importance of small and medium scale enterprises for economic growth industrial development and employment generation at a relatively low capital cost is now well recognised in developing countries. In a small enterprise, the central figure is entrepreneur the person who catalyses resources like risks and manages them so as to establish a viable, sustainable employment generating entity. He/She, therefore, needs to be identified, modified, strengthened and supported. A dearth of such entrepreneurs is a critical problem in many countries, for various reasons entrepreneurs do not spontaneously respond to available business opportunities despite various inducement schemes and programmes to promote and assist them. An effective mechanism is, therefore, sought after to enlarge the pool of local entrepreneurs. In countries where economic activities have been historically dominated by a few families, caste, communities or where educated manpower is not fully utilised, therefore, are problems as well as opportunists for developing a new breed of entrepreneurs from non-business communities and tribes, the first generation entrepreneurs, who can become agents of growth and change.

International development agencies providing assistance to developing countries have often found a lack of mechanism to ensure that the people who receive funds, training services are those who will actively utilised and supply them.

National state level development banks are similarly faced with the problems of a large number of business failure or loan defaults, and one of the key cause is located in the lack of entrepreneurial managerial capabilities in the promoter owner entrepreneurs.

The World Bank has been attracted to new initiatives in small enterprise promotion, particularly in developing countries.

- Services of new entrepreneurs can be diversified and more effectively utilised by tapping those who possess low but developable levels of entrepreneurial traits and managerial capabilities, business information gaps and an initial lack of self-confidence.
- Industrial or commercial work experience comes out to be a by process of entrepreneurs formation. Experience as an employee is an important element in successful self-employment and self-employment experience is equally advantageous to successful entrepreneurship. If, however, the experience inputs are weak or lacking in

an individual, development programme package can be suitably evolved to make up for this in adequacy provided the person has entrepreneurial "raw material" in him/her. Identification and selection of such development developable person is, therefore, crucial to the success of EDPs. It is suggested that education technical education and talent cannot be lumped together and are not fully interchangeable.

- Success in industry has been found to be the result of enterprises and competence rather than financial background of the entrepreneurs. At the same time, potential entrepreneurs as well as financial institutions need protection against ill considered risks. It is the viewpoint that not any one and everyone can be made into a viable business man careful selection and sound training alone may not field satisfactory results unless "need based" financing is provided to the new breed of entrepreneurs who often enterprises and competency to back a sound project. "Having recognised the importance of small enterprises as essential for activating talent entrepreneurial talents in less developed regions and to initiate" a widely differed yet viable process of industrialisation in less developed countries. The nature and characteristics of institutional structure required for small enterprises promotion and development has been one important area of concern of the bank, credit is realised, is only one facet of the problem. The vital facets are identification of viable small business opportunities, provision of required managerial and technical assistance and supply of competent local entrepreneurs to take advantage of business opportunities and set up and run viable industrial or business ventures.

Some of the Indian experiments in the last few years in the small industries development strategy have been well documented in the focus of the achievements of and entrepreneurship development programmes in diverse conditions in India and experiences of the several developing countries suggest that potential entrepreneurship in setting up and running viable small industries is lying untapped in a variety of communities, caste, and occupational groups and this can be effectively brought out and developed by imaginatively conceived and effectively carried out entrepreneurship development programmes (EDPs). A comprehensive training-cum-development programme can seek out and concert non-technical graduates, small traders, and industrial employees, technical and non-technical graduates and unemployed artisans women and sons of farmers, or landless at low cost as fairly well demonstrated by experiences in various parts of India and some developing countries.

- An effective policy tool for opening the doors wide for dormant entrepreneurial potential in an area is breaking the restrictive hold of conventional security dominated banking norms is a necessary pre-condition.
- Entrepreneurial development inputs finances and infra-structure facilities must be treated as necessary package while removing various bottlenecks in the small enterprises (promotion) environment. Such a package calls for sound and timely institutional and organisational arrangement and effective co-ordination among industrial agencies to avoid frustration and delays.
- The key strength of the organisation emanates from the support of banks, select team of trainers and entrepreneurship experts, a committed government support and financial stability EDP's efforts. Innovations cannot be successfully introduced, innovated without creating a suitable funding and International base which can respond to the needs of such venture with a combination of flexibility in policies and attitudes and professionalism of skills.
- The sensibility of performance of the scheme of support institutions character and style is reverted in the conventional banker's approach to the new (first timer) enterprises and subsequent working funds problems which adversely affect the performance of many units.
- The failures of new or small scale units are those which are unable to sell their products in required quantities.

- This is despite state policies to keep the markets bare after reservation of product exclusively, import bans, regulating number of units. The success depends upon the entrepreneurs marketing ability backed by his overall managerial competence in choosing the right product or diversification and competing through in appropriate quality price and sale strategies.
- Inadequate working capital has been another serious problem facing new small enterprises. Long delay in sanctions, non-response to needs at critical times, hardening of attitudes is the first sign of trouble in the unit.
- The first few years are crucial in the environment which is unpredictable, fluctuating, incomplete package of assistance, unless the development finance agencies go full distance in meeting the genuine fund needs and support by effective pre-investment implant counselling.
- Voluntary agencies, especially organisations of entrepreneurs themselves can play an important role in this activity and deserve encouragement and support, while the entrepreneurs can indeed be developed as experience firmly bears out. It is also clear the task is complex. The selection motivation and training expertise, organisational resources for effective EDPs are strengthened and adequate funds are committed to this activity on long-term basis. Developing a team of competent training motivation facilitators to deliver the EDP package deserves priority.
- International development assistance organisations and countries (like India) which are in a position to share experiences and expertise have an important role to play in this field are :

  To train local manpower in selection and motivation techniques and the total EDP training specialists for introduction of entrepreneurship development programme under international development assistance organisation arranging developing country workshops and seminar for.

7

ASIAN CENTRE FOR HUMAN ECOLOGY AT JNU

JNU is much of the experiment as a university where diversity of opinion is a way of life, interrogating received ideas an unceasing quest and intellectual creativity a field for adventure. In the early 1970's, when JNU opened its doors to teachers and students frontier disciplines and newer perspective for old disciplines were brought to the Indian University system. The excellent teacher student ratio at 1:10 the mode of instruction which encourages students to explore their own creativity instead of reproducing received knowledge and of exclusively internal evaluation where also a new experiment on the Indian academic landscape; these have stood the test of time. The very Nehruvian objectives embedded in the founding of the University, National integration, social justice secularism, the democratic way of life, international understanding and a scientific approach to the problems of society has built into it constant and energetic endeavour to renew knowledge through self-questioning.

JNU campus is a microcosm of the Indian nation drawing students from very nook and corner of the country and from every group and stratum of society. To make sure that this is so, annual admission tests are simultaneously held at 37 centers spread across its length groups by reserving 22.5 per cent of seats for them. Overseas students from some 10 per cent of the annual intake and as of now come from 28 countries across the continents. Student's hostels and blocks of faculty residence intersperse one another, underlining the vision of a large Indian family.

Even as classroom teaching and work in the library and the labs have their share in the mode of instruction, personal interaction between students and teachers and among students themselves forms an extremely important and lively medium of generation and transmission of knowledge.

The educational philosophy of the university gets translated into its somewhat unorthodox academic structure. Grounded in a faith in the unity of knowledge, department structure of the conventional universities, preferring instead to bring allied disciplines within a few broad and inclusive parameters called schools under whose interactive ambit are placed the more specialised units called centers. At the Master's level, where most of the schools begin their academic programme, training is largely oriented towards single discipline. At the research level it is the development of research solutions that the going becomes precarious. A series of regional workshops that have focused on linkages between research and development. Certainly the linkages are not near and certainly the processes that give the utilisation of research are complex phenomena that are not completely understood. The problems associated with the practical application of university based research have been taken place in Latin America, North Africa and the Middle East, Eastern and Southern Africa have had a special focus on the problems associated with practical application be brought into line with the real needs of agricultural communities. Understanding the world beyond the research system countries to be one of the principle objective. Parallel efforts will be undertaken to ensure that these technologies are brought to bear within the context of the user themselves and resources will be directed towards increasing the effectiveness of research.

As such research utilisation programme has been designed with three main pillars. The first will be the sub-programme on applied R&D systems. The thirst of this initiative will be to understand better R&D from a system perspective. Failures in the transfer of technology may not be solely attributable to the transfer itself. It is possible that there is a failure in research design, or that there is a system failure. Activities relating to commercialisation of research, within a particular client/market orientation will form the core of this sub-programme. The second sub-programme will focus on institutionalisation capabilities for research application. Here the principle objective will be to built and strengthen the capabilities of research and research supporting institutions to engage in activities related to development for creation of communication of infrastructure network.

The growth of India's agriculture during the past nearly 100 years is closely linked with the researchers done and technologies generated. As a center of higher education and training in agricultural sciences at national and international level :

- To conduct basic and strategic researches with a view of understanding the processes, in all their complexity and to undertake need based research and sustained agricultural productivity in harmony with the environment.
- To serve as a center for academic excellence in the area of postgraduate education and human resource development in agricultural sciences.
- To provide national leadership in agricultural research, extension and technology assessment and transfer by developing new concepts and approaches and serving as a national referral point for quality and standards.
- To develop information systems, add value to information share and serve the information nationally and serve as a data base.

Linkages and Collaborations

For the national leadership in all major agricultural research areas, IARI has close linkages with almost all annual crops and horticrops research institutes, centers, project directorates, coordinated project, as well as few selected institute of ICAR, similar linkages exist for natural resource and socio-economic research institutes collaboration exists with almost all state agricultural universities (SAUs).

Institutional Setting

To supporting and promoting research the task of activising diverse partners in a network of necessity is certainly a challenge for students of organisational design. Indian agricultural research efforts has three institutional basis : SAU system, ICAR and the private industry. The last is yet to carve a niche for itself.

The SAU system consists of 26 universities, at least one in each state. Some states have more than one *e.g.*, Maharashtra and Uttar Pradesh. SAU's have admission capacity of over 1400 undergraduates and nearly 5000 postgraduates while the SAU concept owes much to the US Land Grant College model. The university education commission perhaps described the need rather well cutting for the establishment of rural colleges and universities where no field of human concern should be foreign to them.

The SAU system has about 19,000 scientists engaged in education, research and extension activities. The agricultural university concept was based on the need for a project in which research teaching and extension would be fully integrated and geared towards solving the felt need of the farmers. ICAR and SAU work together supporting each other's activities. The relationships are complex and for our purpose it may be work nothing that SAU is State Government institution while ICAR is central in character.

Agriculture research in India is a mega effort. A vast network of teaching, research and extension groups with such competence employed in solving farmer's problems, the productivity in Indian agriculture should have been better than what it is. A large volume of research solutions to farmer's problems is made available to the community but a few are accepted. The fault may be inadequacy of solution, the transfer mechanism, the limitations of the understanding of the adopters; all or some of them to check the systems of faults and leakage is/are exercise, the other is more fundamental issues.

- Are we conformable in managing such a mega system.
- Does the size itself post its unique problems of reduction to put it bluntly; can the civil serve system foster a culture of innovation where age and seniority are not advantage? It is does not then.

Management of research seems to be as important as the research activity itself in Indian agriculture. These problems are of motivation of achieving excellence and coping with externalities.

Establishment for setting priorities in Agribusiness

As we move towards a learning society every human activity will require contribution from experts and this will place the entire sector of higher education in sharp focus. Although the priorities which are being assigned today to the task of education for all will continue, the country will have to prepare itself to invest more and more on higher education and simultaneously, measures will have to be taken to refine, diversify and up grade higher education and research programmes.

In recent years the pressure of human population, expanding cultivation, developmental projects and increasing urbanisation have been posing a serious threat to our natural plant and animal wealth. Environmentalists all over the world have been great concern about the immediate need not only to preserve, biological diversity but to protect soil health, water resource which are crucial to productive agriculture and the well-being of people.

However, certain developments of relatively recent origin are posing threat to these diverse natural resources. Pressures of human population and urbanisation, the spread of small numbers of high yielding varieties, newer land use patterns and many development projects are constantly squeezing our genetic diversity. It has become absolutely necessary to conserve and augment our genetic resources. In fact conservation is crucial to agricultural improvement. The issue of conservation has become much larger than the concern of environmentalists. Conservation has become central to the productivity of agriculture, which is crucial to the well-being of our welfare the welfare of all of us, in developing world, and developed world alike is internally tied up with the welfare of tropical species and their biological resources. Although the ecological realm of the tropics covers only a limited protein of Earth's land surface, we should not be surprised at this concentration of species with their year round warmth and often with year round moisture the tropics have served as the planet's main power house of evolution. But contrast, the temperate zones and remainder of the planet are relatively deficient in species concentrations genetic resources and biological resources generally. Let us now take a closer look at the contributions made to our material welfare by the wield species through their biological resources in the form of genetic variability. We shall consider four economic sectors in question *viz.* agriculture, medicine, industry and energy.

Agriculture. During the course of human history, people have used around 3,000 plants for food; the great majority of there are tropical plant. Yet the planet contains at least another 75,000 edible plants, again the great majority of them located in the tropics. Of this cornucopia of plant foods only about 150 have ever been cultivated on a food. We are essentially using the same limited number of plant species that have served human kind for millennia. Moreover wild species can contribute to modern agriculture through helping to counter pests and disease. A number of tropical plants produce chemical compounds that repel insects and inhibit their feeding. The toxic compound occur in two main categories; the pyrethrums from chrysanthemum compounds that repel insects or inhibit their feeding and the rotund from roots of rainforest legumes. Many plants and insects notable those of tropical zones have evolved together so we can expect that many other similar insect repelling substances must be available in wild plants of the tropics.

As a science of plant genetics becomes more sophisticated, it can resist the onslaught of pests through innovative techniques. Certain crops can now be sprayed with chemical "messages" that stimulate them to generate insect resistant material. Many other hormone after potential of this sort. While there are only a few primary hormones, whose function is to send specific messages and to affect the plant in various ways, the more numerous secondary hormone trigger specialised responses such as the production of pest resistant chemicals. These critical secondary hormone occur with great abundance and variety in tropical plants.

Commonwealth Institute of Biological Control in Trinidad have total benefits of $ 29.5 million with a continuing annual benefit of some $ 2.2 million. A major further pest control strategy is to control harmful insects through other insect that serve the predators and parasites and thus limits the population.

Medicines. Some further contribution of wild species and their biological sources. The world medicinal; plant resources. As space travel and genetic manipulation open new vistas to mankind, tropical forests still present an intriguing diversity of species that defy comprehension. Our lack of knowledge and understanding of this luxuriant and other vegetation types.

The Earth Ecosystem of India

India is one of the oldest civilisations with scopic variety and rich cultural heritage. It has achieved multifaceted socio-economic progress during the last 53 years of its independence. India has become self-sufficient in agricultural production and is now the tenth industrialised country in the world and to have gone into outer space to conquer nature for the benefit of the people. It covers an area of 3287263 sq. km. extending from the snow covered Himalayan heights to the tropical rain forests of the South. As the seventh largest country in the world, India and the sea, which gives the country a distinct geographical entity branded by the great Himalayas in the North, it stretches south. The Bay of Bengal on the East and Arabian Sea on the West. It has land frontier of about 15,200 km. The total length of the coastline of the man land, Lakshdweep Islands and Andaman Physical background.

For Defining and Integrated Outreach

A form of scholarship that... Involves generating, transmitting, applying, and preserving knowledge for the direct benefit of the external audiences, in ways that are consistent with university and unit missions. Outreach, like other dimensions of the university academic mission is rooted in scholarships for creation, interpretation, communication or use of knowledge. Outreach and teaching, outreach and research, outreach and service, outreach and integration for planning and priority setting to develop a shared vision; resource constraints or limitations to reflect and discuss how can individual faculty, member to serve them best (education, industry, small business, social agencies, health care, government and community organisation with the need driven priority with the thematic and disciplinary objectives and unit strategic plans and external request and problems for the entrepreneurial activities.

For evaluating the quality outreach for impact issues and individuals, sustainability and capacity building university, community relations for the benefit to the university for :

- New opportunities for student learning and professional staff development.
- To innovations in curriculum.
- For the dimensions of the university.
- Mission; to increase cross disciplinary collaborations within the university; in developing outreach potentials and in using the potential to improve the institution's operations and visibility.
- Environment for the development of mathematical techniques for linear and non-linear system. Methodology Air and Noise pollution; application of lasers, Microwaves, and ultra sounds, Bioelectron angelicas to environmental problems. Application of Geology and Geochemistry to rocks, soil and sediment forming process; water pollution and environmental monitoring and management of air.

Conservation of Biological Diversity

For achieving any meaningful success in conservation of biological diversity, we have to have the co-operation of people. They should be sensitised and their needs fulfilled. Special attention need to be given to tropical rain forests which infect are and have been the cradle of biological diversity of wealth are also centers of diversity of our economy plants wild life resources make reasonable contribution in the economy of developing country. The material range from meat of wild animals depending on location, is eaten to the extent of 1 to 70% of the total meat utilised. Skin, fur, horns, bones, wild green vegetables, gums, palms, non-edible and edible oil, fiber, fodder firewood, medicinal and aromatic plants, species, wild relatives of crops and domesticated animals for disease resistant and hardy genes. Many of these small species will continue to be depleted and no amount of legislation and policies can lead to success in saving the species. For instance wild life used for meat can be saved by domesticating some of the species which are smaller, grow faster and can be reared relatively easily. Many of these small species are not known in the conventional animal husbandry but are eaten for the purposes of meat by local people in Africa, Asia and Latin America. Being smaller than goats, sheep, pig, cattle etc., the smaller edible animals, reptiles, fish, birds may constitute our micro livestock. The underlying idea is to relieve pressure on exploitation and ultimate extermination of naturally occurring species the only way to save the animals as also fur trade is to take up sciences-based domestication, production re-introduction, whichsoever is applicable. It would reduce pressure on natural populations, save species and also help the trade. Gene erosion is a common threat to human kind. Biological diversity is inerlinked with protection of environment, plant cover and life support system. The areas of highest biological diversity are the areas of lower agricultural productivity. Development is needed in these areas, but such that it does not ring in a destruction of living wealth and deterioration of environmental. Biological diversity is our common heritage we should realise how important it is for our survival on this planet, and how vulnerable we can be without it.

For achieving any meaningful success in the conservation of biological diversity, production is the best way to conserve them. Besides a set of economic incentive of promotion or production of critical species also need to be developed. This should be backed by appropriate legislation, preventing collection from the wild source and trade of endangered and such other species. Another way to conserve these species is to allot one each of these items to each state for conservation by declaring these as state animals, birds, flowers and trees. A list of 108 species has been prepared by botanical survey of India. Zoological survey of India, National Botanical Research Institutes and National Bureau of Plant Genetic Resources Legislative steps should be taken, making it obligatory on the state to take practical steps for science based conservation and build sizable populations of these species in the relevant conservation area.

All the diversity of forms developed over matures in India has already made systematic system with the establishment of several state and central organisations dealing with agriculture and forestry. This activity was further extended and work on collection and study of crop was carried out on wheat and legumes crops in 1910. However the awareness of genetic resources for crop improvement was created under the imperial council of agricultural research. Considering the national priorities for crop specific and multi-crop exploration was conducted based on richness of genetic diversity/degree rosin. Paralleled to this exploration activity has been the much stronger impact of the plant introduction services of the NBPGR, which is operation for at least 40 years and has now developed linkages with over 70 different countries of the world. The diverse made include agree-horticultural for industrial purposes like Jojoba. With the germplasm build up and the research efforts on crop improvement, need based evaluation studies as ongoing programmes have continued at different activities.

Evaluation and Documentation

Based on the evaluation carried out for various character over the years, the NBPGR has prepared several crop catalogues relating to clusternean, winged bean, cow pea, soybean, lentil etc. Also the NBPGR has prepared several crop inventories on the introduced material of cotton, soybean, linseed, tomato, sunflower, sesame etc. The NGPR has been using the computer facilities of the Indian Agricultural Statistic Research Institute, New Delhi. Apart from the NBPGR the crop-based institutes have also published catalogues on rice, millets, sugarcane and potato collections held by them. With the strengthening of computer network in all such centers, these activities will now be co-ordinate by NGPR.

Water Conservation as a Means of Improving Agriculture

None of the alternatives available for genetic conservation is admittedly perfect. The conventional approaches of germ plasma collection, evaluation and storage in the form of seed have limitations set by the enormity of scale of the exercise and the attendant required input of finance, space and human labour. Similar difficulties are no doubt valid for *in situ* conservation. The techniques involving tissue and cell culture have the limitation of not hang, as yet, perfected procedures and the possibilities of genetic variation occurring during culture and storage. The feasibility of conservation through storage of native DNA is suspected on the ground that an operational gene in the sense of breeding value is much more than the sequence of nucleotides representing the structural gene. What should be the strategy of operation in the light of the doubts and uncertainties expressed genetic erosion will not await the perfection of procedures for conservation and, therefore, it will be prudent to all the available procedures and deploying them in relation to their suitability to different kinds of material targeted for conservation effort.

Designing and Managing Human Resource System

Rural areas are characterised by increasing population pressure, low-level of productivity, widespread poverty and unemployment, institutional and in adequate infrastructure. Three-fourth of population reside in rural areas and heavily depend upon agriculture for their sustenance. Since the very beginning of planning era in India, high priority was accorded to agricultural development as the precondition for improving the welfare of people in rural areas. This was not a choice but compulsion before the policy maker to ensure both agriculture growth and food security. However, all the regions of country could not benefit equally from these strategies which created more regional imbalances and inequality among different sections of the society.

In order to alleviate the poor socio-economic conditions of the farmers, natural resource management through participatory planning in the hill and mountain agro-systern for strengthening family income presently the women are important source for agriculture, animal husbandry and household management. They are the main workforce for agrarian economy in Himalayan region. There were social limitations for women to undertake any innovative step without agreement to male members of the family. This was due to lack of education among women, poor recognition in social system, low status and over exploitation in rural setting and traditions of male dominating society.

Government of India declared the year 2001 as the year of women's empowerment. The year was for merely launched by the Prime Minister in a function held on 4th January, 2001. When five distinguished women were awarded *Stree Shakti* Award for their outstanding contribution to the upliftment and empowerment of women. The department has finalised a calendar of activities on themes like Human Rights for Women, economic empowerment for women, social empowerment of women, women in difficult circumstances, women and technology, women and governance, women and education, women and health and women and media. The main strategy of the women's empowerment programme (IWEP) and expand to create an organisational base for women to come together to analyse, and fulfil their needs through existing programmes of the State and Central Government and to access institutional credit with the view to enhance the capacity of field level function. The department of home science has played an important role in curriculum building, initiation of research activities and designing an appropriate community outreach programmes in the field. The liberal and progressive policies of the university administration have permitted a greater degree of experimentation and innovation than is possible in other institutions of higher learning in India. As a result of this the department of interdisciplinary applied sciences has succeeded in building an interdisciplinary character of the programme in the field.

Recognising the need to prepare competent and dedicated young professionals to meet the challenges of a newly emerging field and to build in them competencies that encompass myriad role of child development professionals to ongoing process of curriculum revisions. The research agenda has changed to keep pace with priorities which emerge as a result of transformations taking place on social, economic and political front in the country, targeting community outreach programmes to vulnerable groups particularly women and children, working in close collaboration with other existing agencies and institutions characterising the field work programme of the centre.

Although many initiatives for women's development and empowerment have been taken tip, still there exists a gap between the goal and directions in the policies, plans programmes and related mechanism on one hand and situational reality of the status of women on the other. However, development of women and children is Center to our national strategy for human resource development.

Women's movement in India is essentially voluntary effort

milliards organised and directed by NGOs. The National Commission for Women is in fact, the result of sustained campaign over a period by NGOs for national level organisations to protect and promote the interest of women. The National Commission for Women has keen associating NGOs in implementing its programmes of holding *Parvarik Mahila Lok Adalat,* legal awareness campaign, public heading and panchayati camps. It has often been brought out that in rural areas particularly in tribal areas the absence of voluntary effort is actually felt. It is an accepted principle of planning that non-government agencies would be associated more and more in implement welfare and development schemes. This can be realised only with the voluntary efforts in areas, which are at present not covered by such institutions. The central social welfare board, the state social advisory board and CAPART play a crucial role in fostering voluntary organisations. The marginalised groups and interior areas require special attention in regard to research of NGOs. It is hoped that incoming year the gaps now in an existence in the spread of NGOs would be adequately filled.

A meet on partnership in action was organised in national consultation with voluntary organisations was organised on 7th March, 1999 at Vigyan Bhawan, New Delhi. Both large and small NGOs, national, regional attended it and grass-root level male and female activities associated with various voluntary organisations working among women. The NCW is placed in an advantage position to tap the strength available with NGO's to bring about attitudinal changes towards women and taking measures for the improvement of their status. NCW identified and key subjects namely :

- Advocacy Education;
- Women empowerment;
- Health in special needs;
- Communication and media;
- Custodian justice;
- Violence against women.

It was recommended that new educational methods must be adopted to ensure that girl child should be given proper education as they remain illiterate especially in villages. No proper attention has been given to adult literacy since independence; no time should be wasted on people exceeding 60 years of age; motivation and right role of the family members relating to children education; more health awareness should be created among women and children. The homework given to school children should be reduced. Improvement of TV programmes, *viz.*, reduction on violence on TV shows more imaginative training programmes; there should be flexibility in school time and schooling techniques; adult functional literacy programme for women instead of adult literacy programmes, educate women about the low cost nutritional food technology.

For creation of vibrant new partnership — to reach the most disadvantaged underserved and the most vulnerable — the young child, the emerging strategy focuses on innovation, fostering decentralised flexible responses to the care of young children, girls and women and reaching through integrated family and community based interventions. It will help in survival, growth and development of socially backward villages and urban slums. The special preference is given to those areas which are predominantly inhabited by vulnerable and weaker sections of the society *i.e.* Scheduled Castes, Scheduled Tribes and low income families found in economically backward areas, drought prone areas and areas in which development of social services requires strengthening. It also believes that these goals are an integral part of all efforts to achieve sustained and sustainable social and economic development that meets needs, ensures well being and protects the natural resources on which all life depends.

The philosophy of human ecology provides the context for developing and serving the individuals and their families from a human environment perspective towards building and sustaining a productive quality of life. The scientific expertise, infrastructure facilities and research needs to be strengthened by bonding the relationship between research and educational institutions and industry in a symbiotic mode of conversion of research findings into products, processes and technologies for human resource development.

Interdisciplinary knowledge and skills to deal with the problems through vigorous research, active extension education programme will outreach the community. This will help them to meet the challenges imposed by the new economic and social order and improve their quality of life. The concept and view-points developed here will be helpful in framing educational strategies in higher education and research needs in the field of human ecology.

The key subjects namely Advocacy, Education, Women Employment, Health and Nutrition, Women in Special Needs, communication etc. to build up capability, to able to ensure that the delivery arrangements for programmes should be efficient and they should reach those for whom they are meant. The voluntary organisations have emerged as most viable institutional framework to serve as catalysts for women's development. It would be very important for the Government to invest funds in the women's movement and provide political support to it.

Our networking will not only liberate; built blossom the long suppressed *Stree Shakti* of our people. In education new educational methods must be adopted to ensure that the girl child should be given proper education, as they remain illiterate especially in villages and slum dwelling women. No proper attention was given to adult literacy therefore, there should be functional literacy programme for women instead of Adult Literacy programmes. Empowerment should be generated among the women. It cannot be transferred from outside. Mobilisation of women folk, mobilisation of resources and participation of women in policy formulation in any perspective of society should not only involve the Government and Non-Government sectors but also equally involve the men. This should be taken as holistic approach. Empowerment can only be done economic upliftment. Though it is criticised that 33% participation for women in Panchayati Raj is in effective as they have not acquired proper skill but in due course of next 5 years they will definitely acquire it. Collaboration with NGOs to bring economic development and empowerment of women by individual, behaviour and counselling or NGOs. It should also include some men in their managing body. It was stressed that some public inquiry should be conducted by NCW and it should monitor the land distribution agricultural wages. Existing policy implementation labour laws etc.

8

Research Methodologies

Religion in India

The main currents in Indian Sociology intended to be an outlet for ongoing social science research on the Indian sub-continent, explores fresh areas of research, deal with various aspects of religion. Primarily Hinduism so as to explains its role in the changing social structure of Indian society.

Conducting respectable research on religion in the Indian sub-continent has not been the easiest task for the social scientists. All major religions with diversified sectarian ideologies along with tribal belief system, reformist schools and cults founded by pragmatist gurus pervade the society. Considering the immensity of the subject and the myriad groups involved, the volume is restricted to Hinduism and its followers with the hope that a complementary volume on the studies of non-Hindu groups will present a balanced coverage of the subject.

Hinduism did not escape the attention of great pioneers of social sciences like Emile Durkeim. Commentaries, discourses and interpretive analyses, through common in the past, were generally either philosophical mythological propagate or normative.

Systematic studies during the nineteenth century were conducted for the British Empire to help understand the cultures of the colony for effective administration. Among the surveys, personal diaries, travel accounts, memories, and of course ethnographic reports, the most significant were *The Imperial Gazetteers of India,* 112 volumes of which were prepared in 1869 under the guidance of Sir William Hunter and the sponsorship of the Government of India. Volume VI dealing with Hinduism, "distills into one volume the essence of the whole.... Hinduism is the religious and social nexus of the Indian races" (1886 : X). This view was further strengthened by Sir Monier Williams (1891) who said that the most vital quality of Hinduism is its receptivity and all-comprehensiveness. He adds that it does not oppose the progress of other systems, for it has no difficulty in including all other religions within its all-embracing arms and ever-widening fold. Its very strength lies in its infinite adaptability to the infinite diversity of human characters and human tendencies.

Perhaps even more than other religions. Hinduism is a complex religion and its myriad forms and practices have created methodological problems for social scientists attempting to report systematically the very spirit of the religion and the determinants of actual behaviour of its adherents, their hopes and aspirations.

Mandelbaum (1964 : 5-20; 1966 : 1174-1191) one of the most ardent assayers of Indian society, succinctly puts forward the social scientist's view of religion and says (it) "takes in the whole of a people's belief and practice toward the supernatural, that is toward their concepts of all that is beyond man's own power and mundane knowledge." This perspective not only takes into account the scriptural explanations of religion but goes beyond to look at behavioural facts and religious activity in total context.

The varieties one finds in local or regional Hinduism have sometimes been problematic in attempts to dismiss Indian Hinduism. Marriott attempted to explain this relationship through the concepts of universalisation and parochialisation (1955). While Marriott addressed the linkages at local, regional and national levels, Mandelbaum looked at the functional aspects. Reviewing a wide spectrum of studies on religion in South Asia, he formulated two complementary complexes. The first one he calls "transcendental" which is ultimate, supernal derived form the Sanskrit texts and relates to system maintenance, long-term welfare and ultimate goals; universal gods and annual cyclical rites, and its derived from the knowledge of higher priests who are acknowledged to be the reservoirs of religious wisdom and learning. The other is the "pragmatic" complex composed of lesser, regional or local deities of various types, including the spirits of the dead, the regulators of disease, epidemic, natural disasters and community guardians. In contrast to the transcendental complex, this complex includes personal and proximate goals of individual welfare, the rites related to local deities and spirits and the ritual performers who are not necessarily equipped with the knowledge of Sanskritic Hinduism.

These discourses, although significant, have not yet satisfied the scientific inquisitiveness of scholars, some of whom believe that Hinduism provides a uniquely different environment for the socialisation process of the individual. Therefore, the pre-eminence of religion in the make-up of the Hindu personality has recently been a major issue in the domain of social sciences in that Western scholars have for the first time acknowledge that Western social science has yet to develop new concepts, and possibly a new methodology, to study Indian society. Dumont's (1970) attempt to understand Indian society on its own terms and Marriott's (1974) ethnomethodology are the major attempts to link the person to the culture and to show how social interactions, help foster unique qualities among persons within the system. Is the Hindu personality more socio-centric than egocentric? Are there any basic differences in the world view of the Hindu person compared to Christian or Muslim? Recent research may be able to provide some answers to these questions.

In general, Hinduism is an amorphous body of beliefs, philosophies, worship practices and codes of conduct. The very nature of Hinduism leads to a greater tolerance of other religious thoughts among its adherents as they tend to believe that highest divine powers complement each other for the well-being of the world and mankind. Rarely are any religious ideas considered to be finally irreconcilable. Often, other religious ideologies, though sometimes claiming to be diametrically opposite, may find that most of their basic elements have a place in Hinduism. The qualities of resilience, absorption and respect for alternative ways of reaching the same goals, perhaps, are the characteristics which have generated vitality in Hinduism from millennium. Arvind Sharma attempts to define population and historical that there really is a common cultural base for society in India. During the past three decades, advances in social science research have been significant; however, Preston remarks that extensive survey research is needed as case studies alone will not help bridge the gaps in our knowledge of the Indian mosaic.

Social movements in indigenous societies usually originate against potential real or imaginary dangers perceived in society. It is hardly likely that they intent to combat undesirable and threatening changes emanating from abominable forces believed to be disruptive for a community will go without a counter attempt. Such efforts form a movement which demands a common base which is readily available in a common belief system. Recent episodes in the international field further suggest that religion is one of the primordial forces that unite as well as to divide nation(s).

John MacDougall discusses the bases of two social movements among tribal communities in the state of Bihar. He finds that Christian and non-Christian tribals both used, although in different ways, religion as a stanchion to revitalise their society and strengthen the movement to challenge the authority of the British during the nineteenth century.

In the other paper on the tribal society, the Toda of the Nilgiri Hills in South India, Paul Hockings and Anthony Walker report a rare account of the "dry" funeral involving complex symbolism and its relevance to the various segments of the Toda community.

Although there are numerous castes and communities whose historical antecedents reflect religious conversions through various methods, their current mixed religious affiliations are rarely challenged, yet they face identity problems. As elsewhere in the sub-continent, groups embracing a new religion did not actually change many of their basic values and characteristics. This has been equally true among the Hindu groups who proselytised to various sects of Islam during the Islamic rule. Most retained their caste-like character and certain customary practices of Hindu origin.

The process of assimilation though started centuries ago, in some cases was never consummated. One such case involves the Malkana community in the state of Uttar Pradesh. The problems of identity among the Malkanas are enormous due to their dual religious affiliations, identifications, and family names. Due to selective retention of religious traits, the Malkanas are neither fully accepted as Hindus nor Muslims. Although in the village environment, identities usually do not become a matter of concern, they turn into a cause for schism in an urban setting where the institutional aspects of religious groups are more pronounced.

Addressing a related question, Yogendra K. Malik discusses his findings from the northern states of Punjab, Haryana and Delhi. He asserts that in many cases, scholars started their research with wrong assumptions. For example, in a developing society, one finds the elements of weak legitimacy, languid national identification and a strong loyalty to religious leaders. His extended analysis of the Sikh youths shows that their socio-political values are no longer determined by the traditional agents of political socialisation, such as religious leaders, but by secular, considerations and exigencies of the political system. The link between the youths socio-economic origins and their exposure to modern and secular values is clearly evident. To qualify his observations, he points out that as it happens in any democratic system, people politically profit by using various means of identifying groups to their advantage. During the past two decades, several studies have resulted in mixed findings, including those suggesting that caste, religion and region have gained prominence in the democratic process. As elsewhere in firmly established democracies, politicians look for their bases and tailor their speeches and approaches accordingly. It is no wonder that already known aggregations and cleavages serve as ready-made bases for politicians.

The last two papers deal with religion as it is found among Indian communities outside India and a comparative analysis of the Indian and American world views. Michael V. Angrosino's paper covers the lands of Southeast Asia, the Pacific Islands, Africa and the Caribbean where Hindu and Muslim communities of India have settled, some for generations. The extensive description and analysis suggests that, although the reasons for migration have been different in each country, at various points in history, the sharing of common religious values, rites and ceremonies has played a vital role in maintaining Indian identity. He points out that pride in their ancient culture and family life grounded in philosophical cosmology and religious-ethical prescripts not only governed social life, but also symbolically validated ethnic superiority.

Grace Cairns shows that though the processes of urbanisation, industrialisation and modernisation supposedly have caused a decline of religion in other societies, the highest values of Hinduism *dharma* and *moksa* continue to significantly influence individuals lives. Interestingly, the reformers, the modern interpreters of religion, the modern gurus, and the freedom fighters in pre-independent India, all found the basic tenets of Hinduism as the sanctuary for their new goals. Cairns clearly lays out the processes involved from vedic times to the evolution of the organisations, such as Ram Krishna Mission and International Society of Krsna Consciousness. Among the many ways the structure and functions of Indian society is interpreted. Louis Dumont's ideas are unique as he persuasively argues that the individual in Indian Hindus society is not governed by independent thought and action commonly found in the West. His major thesis, clearly expounded in Homo Hierarchicus (1970), has been criticised by Marvin Davis. Based upon his work among the Bengali Hindus. Davis suggests that Dumont did not see much beyond what he wanted to see and failed to discern a continuity in structural opposites.

International processes which help in maintaining the system and resolving conflicts. Susan S. Wadley examines the enormous influence the vernacular religious press and the increasing literacy rate has been exerting in strengthening oral traditions is South Asia. Suzanne Hanchett reports from the south Indian state of Karnataka on the relationship between festivals and political processes, suggesting that festivals are events of power manifestations, negotiations and ongoing conflict laying out the framework for dynamics of the changing system. Another paper by Janet Benson analyses one of the major festivals of Sanskritic tradition, Dasara in the Telangana region of Andhra Pradesh. She finds that religion not only represents the metaphors for social relations but in fact shapers and regulates the social order. The interplay of variable of rank, prestige, caste, class and power are recurrently displayed and legitimised. Here we find that religion not only relates the individual to the supernatural, but also to his social environment and to a great extent affects psychological social and cultural processes.

Ruling out the possibilities of the blending of cultures such as European, Indian and African, he concludes that the links with the "great tradition" of Hinduism present an evolutionary process within Hinduism more than cross-cultural diffusion. Migrations have led "little traditions" of Hinduism which are reinforced by the preachings and discourses of visiting teachers of religion. The new lifestyles do not seem to disrupt the significant role religion plays in personal need, satisfaction and identity maintenance.

This proposition is further advanced and clarified by Paul G. Hiebert through a comparative study of American and Indian world views which are largely influenced by religious philosophies, ideas about the fundamental nature of universe and explanations of reality. His models are American middle-class, followers of Greek and Judeo-Christian ideas and the Hindus of Andhra Pradesh.

As social scientists conceive religion, they believe religion is based upon a set of assumptions. These assumptions are the ones which determine the character of religion's values and norms or even rituals and customary practices. Incidentally, most religions share the same existential problems of explaining the unexplainable and providing meaning to existence itself. Nevertheless, these assumptions to a great extent determine the day-to-day behaviour patterns of the people. For the social scientists, it is not so much relevant to understand why these assumptions came into being, but how they relate to the actions of the people. Differences in assumptions may be crucial to the understanding of both social systems.

Each of these papers reveal that religion continues to be paramount force in the lives of the people. There is no clear indication that there has been any decline in religious rites or ceremonies. Incidentally, the influence of modernisation is overshadowed by religious rites and values. It is generally palatable to the Indian mind to assimilate the new in the old order of things. With rare exceptions, we find that religion has been used in technological advancement, the practice of modern medicine or aggrandizing political goals. Religion continues to provided a support system for new structures and new ideologies. Any anticipated changes thus should be viewed rather in the larger context of the religious system without which any generalisations about Indian society would be either erroneous or incomplete.

METHODOLOGY

To be concerned with man in the social world it is necessary to explore the reality which underlines and characterises 'man', the 'social', and 'world'. Philosophy is inescapable for the social scientist who seeks clarity and rigour in work, who takes the term 'methodology' seriously. It is not merely a question of interest in the logic of scientific enquiry, a matter of paying attention to the status of propositions, models and laws. For paying attention to the phenomena of the social world, symbolism and meaning of social action; men acting in the context of an inter-subjective reality, shared and sustained by temporal beings aware of themselves no less than each other.

A crucial aspect of methodology relates to the relationship between the observer and the observed. A host of philosophical problems concerning the nature of the subject, subjectivity, construction of the subject, and in consequence textual construction of subject, but also of consciously trying to enter into the world of meanings of the subjects under study. The phenomenological critique of sociology has stressed the fact that data gathering is a social process and needs to be understood as such.

With regards to various approaches to methodology that are open today. For anyone looking back over the preceding survey, the most striking fact remains the dominance of the positivist tradition in methodological writing and its influence in the form of the desire for adoption of the 'methods of science'. This has meant, in practice, the devising of a particular mode of enquiry stressing the importance of concepts, hypotheses linked to defined variables, empirical testing and formal analysis. In conjunction with this process social statistics have been used to aid both in the construction of the empirical test and in the analysis and interpretation of the results. The other themes that emerge from the alternative tradition which has been based in social anthropology, village and community studies.

While exploring the themes attempt has been made to identify their problems and implications. These are issues related not simply to techniques of research, but also to the questions of meaning in an inter-subjective world, justification of applying statistical logic and experimental method to a post-positivist universe, nations of objectivity, science reality and realism, methodological implications of construction of subject and test and so on. The general tenor of methodological texts has been to provide a stereotyped account of 'scientific' method, research design, hypotheses, observation, interview in an uncritical fashion, following the model of scientism and without in any way incorporating the current views of these subjects.

The problem that arises in this kind of work concerns the nature of conceptual categories to be employed and this kind of research method has the inherent danger of allowing the social scientist of negate his vital function of making sense if the data, of consciously constructing the problem rather than expecting it to emerge at some later stage.

Our aim is to go beyond mere descriptions of various methods, to examine the theory, logic and philosophy of the methods and to explore their implications. The questions of methodology proper are linked both with the theoretical issues related to philosophy and to applied issues associated with various mathematical and quantitative techniques. At an abstract level, questions of method and questions of theory overlap and are often identical. They are related to various theoretical positions like positivism, phenomenology, structuralism, post-structuralism and so on. The first task, therefore, is to look at some of the main elements in methodology and how they have been put into operation in the research process.

In the typical methodology text, the 'traditional method' is contrasted with the 'scientific method', and described as something prohibiting proper research. For instance, this is how the features of the traditional methods have been described : (i) methodological *ad hocism,* (ii) eminence of subjective perception, (iii) ethical preoccupation rooted in the refusal to distinguish methodologically between facts and values, between 'is' and 'ought', (iv) the dominance of quality against quantity, and (v) the greater use of intuition and imagination, which are regarded as essentially disruptive of modern scientific methodology, are the constituents of the traditional method. The conventional methodology is based on the notion of 'scientific objectivity', which is related to bias in social science. The problem of scientific objectivity is still worsened because of the bias of the social scientists governed by conscious or unconscious political and social attitudes and values.

Fieldwork Methodology

Fieldwork approach in principle rejects the positivist assumption that descriptive concepts are simply a first stage towards the test of explanatory hypotheses. The concern for interpretation of culture and meaning and interest in practices through which social relations become visible from the basis of fieldwork methods. In contrast with the sharp distinction between concepts and explanatory hypotheses within the positivist approach, in the interactions and interpretative field work method 'understanding' the actions and interactions of respondents by virtue of grasping and comprehending the culturally appropriate concept through which the people conduct their social life, is the way in which the explanation is achieved. In the field work method the focus is on naturally occurring situation, the encounters that people have in normal course of events. In this way, it is hoped to produce descriptions and explanations appropriate to the ways in which people actually behave.

There is no dearth of books on research methodology, and almost all aim teaching the novice statistical methods, survey research and its history, drawing up different kinds of questionnaires, interviewing, opinion polls, content analysis and market surveys. Scientific method in the social sciences, in which the need is stressed for the clear formulation of hypotheses to be tested in the inquiry, appropriate research design, the elimination of bias in the collection and interpretation data etc. Throughout, the assumption appears that the social sciences are of the same character as the natural sciences. The new developments in social anthropology, in their view, have blurred the distinction between anthropology and sociology, particularly so in India. Though in methodological terms a conceptual distinction can be made between ethnographic field work and participant observation in which the former involves a recognition of the interdependence of theoretical and methodological issues and the latter relates more to description in terms of participation which often trivialise the issue involved to one meter technique. Like the discussions of how to obtain access and/or to remain concealed and so on — in Indian field work literature no such distinction is made.

The account of field work cover a very wide range of field situations and subjects. The varied locale and subjects cover slums, factories, trade union, industrial cooperatives, college students, school, hospital, minorities, nomads, small towns, and of course, villages. The accounts also point out some important aspects of fieldwork, which are not generally discussed in the available textbooks. However, the field work account argue that hypotheses cannot be formulated in a vacuum, without knowledge of the field. Moreover, the field workers cannot anticipate the developments in the field, which will inevitably guide the course of his investigation. Hypotheses formed without regard to these considerations may turn out to be trivial if bot banal. Almost no contribution to this volume has been guided by hypotheses, and some confess that their theorizing was only post facts. What most field work do is to go to the field with a founding in the theory of the discipline, especially in the sub-area of their interest, and with as much knowledge of the region as can be derived from secondary material. The field situation offers him opportunities to reformulate existing concepts and invest them with new meaning.

Survey Methodology

One of the commonly used methods in empirical social science research in India is the survey method. The theme concern how survey research must be designed and conducted as an institution of first accounting of the great variability within them, and second becoming more relevant to the social cultural environment within which measurement occurs. Subordinated to these key issues, the need for using complementary research methods to support and enrich surveys; the need for pre-study exploration prior to conducting the full study; the significance of human errors; the act non-randomly to bias and thereby weaken the survey methodology; the need to study non-random sources and types of biases and describe these methodological findings so that study can be replicated and socio-economic feasibility of implementing results be known.

Inductive Methodology

Provide the basis for research methods can be described as deductive monological and inductive-statistical system attempts to provide a basis for unification of the sciences and also embodies certain specific features of positivism.

Systems of Social Indicators

With the increasing demand for the present-day literature dealing with social policy for the development of systems of social indicators various methodologies have been geared for developing appropriate measures.

Mathematics, Computers and Formal Analysis

A body of theory in sociology and anthropology that does not call for mathematical treatment, for any theory that conceives of society or culture as a system is committed to the notion of interrelationship of parts, and all conceptions of the cultural consensus and conformity on the one hand and degree of complexity on the other can involve nations of formal analysis, of measure and scalar arrangement as their abstract representations.

RESEARCH DESIGNED AND METHODOLOGY

Catalyst Role

- Acting as focal agency in the University system.
- Serving as technical resource center for universities community activities.
- Assessing the need of the community.
- Developing curriculum and multidisciplinary University programmes, involving faculty students from various disciplines.
- Stimulating other departments of the University to undertake similar activities with respect to their discipline for continuing education of graduates, extension education and field outreach.

Teaching, Training and Research for Socially Disadvantaged Group

- Under the framework of women participation in management, integration of services, at an institution to provide feedback guide implementation, to improve quality of life in farm, off-farm and non-farm sector for human resource development.
- Development with special reference to tribal women through the strategy which include social empowerment, education and development.

A project proposal entitled "Research and Training in on-farm, off-farm and non-farm sectors for human resources development with special reference to tribal women".

- Teaching programmes/course in adult and continuing education :
- Conducting need based continuing education courses not undertaken by other departments of the university.
- Conducting activities/programmes for upgrading knowledge skills or acquiring new skills, short-term diploma certificates programmes.
- Training short-term aimed at various groups is skill income generation, entrepreneurship development, family life education. Para legal training for grass-root workers/activities.
- Community outreach activities — responsibility to reach out to society, specific disadvantaged groups, *e.g.*, bonded labour, child workers, urban slum and villages.

Programmes

Vocational/Career oriented programmes.

Empowering people at different socio-economic and educational levels with marketable vocational skills, which are appropriate.

- Participation in literacy, post literacy, continuing education under the ground level model.
- Evaluation as per university norms.

Collaborating and Networking

- Networking with government departments and other agencies of the country.

- International collaborations to implement the programmes for higher education and to draw the programmes of collaboration by the participating institute — for linking higher education in the priority areas of environment, population education and science and technology and gender issues.

Documentation particularly of Field Experience to Enrich Theory and dissemination of Information.

- Preparing material.
- Disseminating information.
- Use of Computers and Internet-Publications.

Monitoring and Evaluation

- Monitoring and Evaluating the Department's activities as well as ongoing extension in the university.
- Reporting to Nodal Center/University.

RESEARCH METHODOLOGY AND LINKAGES

(To Impart Relevant Education and Training Programmes in Income Generating and Development Activities and Linking with Home Science Curriculum)

Introduction

In social sciences methodology is taken to be a discipline, bordering on philosophy, whose function is to recommend and examine the methods which should be used to produce valid knowledge. Methodology lays down procedures to be used in generation of valid knowledge and these procedures are justified or criticised by means of philosophical arguments.

The project aim is to go beyond, mere descriptions of various methods, to examine the theory, logic and philosophy of the methods and to explore their implications. The questions of methodology proper, are linked both with the theoretical issues related to philosophy and to applied issues associated with various theoretical positions like positivism, phenomenology, structuralism, post-structuralism and so on.

However, there are critiques of objective research and positivist methodology in Social Sciences. The widespread acceptance of conventional methodology has been seriously questioned both on theoretical and political grounds by some Indian scholars.

In the past 15 years, developing countries' concerns about persistent rural poverty and malnutrition. Unsustainable agricultural production, lack of food self-sufficiency, and swelling urban populations have become increasingly acute. In response, many governments have undertaken major initiatives to help resource — poor farmer households improve their living standards and productive capacity. These farmers make up not the majority of producers, but also a major share of the world's poorest and most vulnerable people.

The challenges for meeting the needs of resource poor farmers is to develop a greater capacity to facilitate effective interaction between researchers, technology transfer workers and resource poor farmers. This requires :

- Shifts in research policies and priorities.
- Changes in the organisation and management of research and technology transfer agencies.
- The development of strong links between these agencies and resource poor farmers.

THE LINKAGES
RESEARCH AND TECHNOLOGY TRANSFER LINKS

(Research Methodology to Impart Relevant Education and Training)

The educational institution have clear goals that are compatible with national policies and defined channels which users can make their demands known, it is still important to recognise that the tasks of producing and transferring relevant technology are carried out by two groups of people — researchers and technology transfer workers — and that these groups differ in attitude, interests and training.

There are series of tools and mechanisms which management can use to promote links among researchers and between researchers and technology transfer workers. In my view, committees, liaison persons and coordinating units often have a difficult time creating effective links, unless they have power and authority or control a budget.

We have to be very precise about what kind of links are needed, at what level they should be established, and what goals we hope to achieve. When we need so many linkage mechanisms that an organisation becomes excessively complex, we have to start thinking about reorganisation rather than creating more position or units in the organisation.

The goal of developing links to build bridges between researchers and extension workers should not make us forget that specialisation and division of labour have contributed to growth throughout history. Sometimes a bit of competition between actors will lead to improved performance by each of them, as long as the social institutional goals are clear for everyone.

Home Economics to Human Ecology

How the concept and philosophy of human ecology can be implemented through teaching research and outreach? From its inception human ecology has focused human being relationship to the material goods and social and economic system necessary for living. Over the courses of its history, human ecology will seek and applied knowledge from many sources and disciplines helpful to individual and family in meeting everyday needs and addressing significant human environment problems. The focus has been and continues to be on the activities of everyday life associated with meeting recurring sustenance needs for food, clothing and shelter and human development and relationships within various environments such as the family work place, market systems and other community settings.

Historically, a major focus has been on the micro environment — home and society. As society technology, and home science institutions have evolved, increasing attention has been given to other settings in which needs are met and to interdependence within and across family, home, work, business and institutional environment. The history provides a sense of the holistic and ecological nature of the home science colleges and shows the ways in which inter-related functions and components are connected. The project may not provide specifies about academic majors research projects or other scholarly activities, not details of the far-reaching extent of community service, cooperative extension and other outreach programmes of the college.

This project is an attempt to capture the spirit of Home Science to Human Ecology its development, growth and challenges for new social demands, evolving budget and other opportunities and constraints.

The Purpose of Home Ecology/Human Ecology

Concern for the well-beings as and families. The field is defined as the study of human beings as an integrated whole, his man environment and the interaction between them. The focus has been the interaction of humans with their environment, emphasising.

In-depth study of the reciprocal effects that take place physically, socially, and aesthetically. Home economics had originally begun as a unified field for the application of science to everyday life in the home, family and community but over the years had developed into specialised fields and professions. This development reflected scientific, technological, socio-economic and political changes as well as changes in the family. Including greater attention to the development, selection and use of consumer goods by families and individuals.

Emergence of an Ecological Perspective

The term ecology and ecological soon became art of the many lively discussions. Concern increased over the effects of human actions on the environment, including widespread pollution, which endangered human health. Actions that threatened survival of various species of plants and animals created renewed interest in ecology, the science concerned with interrelationships between organisms and the environment, both organic and inorganic. The term ecosystem came into use, and the ecological roots of home economics were rediscovered. "Oekology be henceforth the science of (our) normal lives.... The worthiest of all the applied sciences which teaches the principles on which to found...healthy and happy life. Thus home ecology as a means to improve people's lives and environment. This should be done in home and family where basic needs are met and human character is formed. Primarily responsible for the activities of home and family, women are the major focus of home ecology. However, this did not become the name of the field, instead home economics was the name adopted. The family as an ecosystem provided a philosophical basis for human ecology as it applied to the family and to the home economics.

9

POSTGRADUATE AND POST EXPERIENCE PROGRAMMES

- Criminology
- Social Policy
- Psychology
- Geography and Environment Management
- Social Science Research Methods
- Psychoanalysis
- Social Work
- Research Degrees

Social Science Research is based on a shared approach to the exchange of ideas, expertise and skills between students and staff. Our flexible modular structure means that you have the opportunity to build a programme which meets your specific requirements.

JNU has the facilities, and our staff the special skills, required to help both home and international students to succeed in their studies.

Social Science

1. Graduate studies in the field of Social Science.
2. The Graduate School.
3. So why study for a taught masters degree in the school of Social Science.

Programmes

4. Multidisciplinary Programme.
5. M.Sc. PG Dip. Research Methods in Social Science.
6. M.A. Criminology.

7. M.A. Criminology (distance learning mode).
8. M.A. Social Policy.
9. M.Sc. Psychology and Health.
10. M.A. PG Dip. Social Work (Dip SW)
    M.A. Social Work 'Top Up'
    Employment Route MA/PG Dip. Social Work.
11. M.A. PG Dip. Youth Justice, Probation and Applied Criminology.
12. M.A. Psychoanalysis.
13. M.A. M.Sc. Sustainable Environmental Management.
14. M.A. M.Sc. Sustainable Environment Management (Geography and Environmental Management).
15. M.A. M.Sc. Sustainable Environmental Management (Water Management).
16. M.A. Sustainable Rural Development.
17. M.Sc. Arboriculture and Community Forest Management (Joint collaboration with Capel Manor College).

Other Programmes offered at the Enfield Campus

18. PG Dip. M.A. M.Sc. Interprofessional Health Care.
19. M.Sc. PG Dip. Mental Health Interventions.
20. M.A. Local Economic Development.
21. Opportunities for Research.
22. Staff interests.

POSTGRADUATE AND POST EXPERIENCE PROGRAMMES FOR MAKING A DIFFERENCE IN SOCIETY

Empowering people at different socio-economic and educational level with marketable vocational skills which are appropriate to market demands both in rural and urban areas. For students who have undergone general education causes in continuing education could offer special for those already in jobs to update the skills or acquire qualification for career advancement. Need based relevant education programmes to fill in the gaps in their intellectual growth, enhance professional and technical competence and promote undertaking of contemporary issues.

Stimulating to other departments of the university to undertake similar activities in assessing the need of the community with respect of their discipline for continuing education of graduates extension education and field outreach. Developing curriculum and multidisciplinary programmes involving faculty, students, from various disciplines. This can help in developing the discipline through academic programme and courses in criminology, social policy, psychology, geography, Environment management, social science research methods, psycho analysis, social work and research degrees committed to high quality learning and research.

AT CENTRE FOR OVERVIEW

Transformation of Home Science to Human Ecology

Transformation from Home Science to Human Ecology for community empowerment women's education, today especially in rural areas must form the main agenda of Government. Professional orientation to curriculum, enhancing professionals for making graduates self-reliant, the professional courses would provide under spectrum of experience to graduating students have sufficient weightage have to be given to make students professions and befitted to face with vigour and confidence the challenges of society for self and family, scientifically and technically in the most befitting ways.

Transformative directions reflect the careful and informative thinking. It is the result of our collaborate process with national and international home service education system can integrate and provide a holistic perspective of home science colleges.

GRADUATE STUDIES IN THE FIELD OF SOCIAL SCIENCE

Social Science supports and organises teaching and research activity. This includes taught Masters programmes and M.Phil. Ph.D. training with students following a variety of courses part-time, full-time or by distance learning. Teaching and research training are underpinned by a lively research culture involving both staff and students focused around conferences, seminars, forums and other activities, together with lectures from invited speakers.

Social Science research at Middlesex is recognised as work of national and international standards of excellence as evidenced in ratings awarded in Research Assessment Exercises. All subjects share a distinctive Middlesex mix of critical theoretical study, policy analysis and applied work. The key subject areas are social policy, criminology, social work, psychoanalysis, sociology, gender and racial equality studies, geography, environmental management and psychology. The focus of work ranges from international and comparative accounts to studies of national and local issues. Key topics include crime, domestic violence, prostitution, prisons; flood hazard, coastal management, risk and environmental sustainability, palliative care, work with refugees, care for vulnerable groups; racial equality studies, equal pay and equal opportunities, gender and womens studies; changing welfare, public sector management, poverty and social exclusion, social problems and social policy; cognitive and health psychology; psychoanalysis and psychoanalytic theory and much more.

Involvement in a Learning Community

A postgraduate community which provides its members with support and stimulation. Academic staff facilitate learning in a variety of ways — lectures, group work, individual research and tutorials — which are designed to help you become self-sufficient in organising your own study and development. Academic staff involved in postgraduate work are active in research and consultancy, producing publications regularly. Research is disseminated in the teaching.

Research Centres for Criminology, Flood Hazard Research, Gender Research for Psychoanalysis for Environment and Safety Management for Business and Social Policy Research for Racial Equality Studies.

M.A. Psychosynthesis Psychotherapy

PG Dip. Psychosynthesis Therapeutic Counselling

PG Dip. Applied Psychosynthesis.

All students have access to powerful mainframe computer facilities and a large number of high specification micro-computers. These are networked to provide a wide range of software, communications and print services. There is a dedicated postgraduate computing laboratory and a comfortably equipped, open access language centre provides audio and video teaching materials as well as strong tutorial support.

Quality

In addition to meeting the standards set by government, through the quality Assurance Agency, its own robust system for assuring quality. *Inter alia* this involves experts drawn from other universities who participate in the development of new programmes as well as reviewing existing programmes.

The commitment to quality, as well as continuous improvement to the programmes stemming from the research and publications of academic staff, can mean that changes to the structure and curriculum may be introduced following the publication of the programme details in this brochure. We will make every effort to notify applicants of any approved changes.

- M.A. Development in Practice.
- M.A. M.Sc. Countryside Management and other land based professions.
- M.A. Criminology.
- M.A. Racial Equality Studies.
- M.A. Social Work Studies.

Programmes will be added to like currently available for new developments.

Learning Facilities

Facilities to back up the study programme are extensive. The university library has over 400,000 books, 2,200 periodicals and an extensive collection of videos, slides, records and other materials. The collection is computer indexed to allow easy access to material throughout the library. Facilities for freely accessing data via CDROMs, on-line and stand alone PCs and the Internet are increasing and are available both in the main library and the Postgraduate Reading Room.

STUDY FOR A TAUGHT MASTERS DEGREE IN THE SCHOOL OF SOCIAL SCIENCE

Social Science has a nationally recognised range of high quality postgraduate programmes. Our programmes are designed to meet the needs of many different types of students and to offer different modes of learning. It is possible to switch between part-time and full-time study (where appropriate), to interrupt study and to study at your own pace while accumulating credit points for an award. A distance learning mode is available for the M.A. Criminology. It is possible, with the M.A. Multidisciplinary, to combine two programmes. Also, you might not wish to take a full masters programme but perhaps one or two modules, either because they interest you or support a career change. If you satisfy the enter requirements for masters level study, you can enrol as an associate student. Postgraduate and research students have the opportunity to spend time overseas to study, conduct research or gain work experience.

M.A. MULTIDISCIPLINARY PROGRAMME

Mode. One year full-time, or two years part-time depending on content.

This is a masters programme which allows students the flexibility of combing modules from two postgraduate programmes.

Programme Structure

Students combine postgraduate programmes either solely from within the School of Social Science or from other Schools within the University to form a Major subject with a Minor or a joint Major. For example, using programmes from within the School, a Joint Major in Criminology and Psychoanalysis could be taken or a Major in Social Policy could be combined with a Minor in Human Resource Management or Management Practice offered by the Middlesex University Business School. The name of the final award will reflect the combination of subjects, *e.g.*, M.A. in Criminology and Psychoanalysis or an M.A. in Social Policy with Human Resource Management. The programme leader will give individually tailored advice to help students plan their Multidisciplinary Programme both in advance of starting the M.A. and during the period of study.

By combining Research Methods with a subject area, *e.g.*, Social Policy and Research Methods or Criminology and Research Methods, not only will students gain a strong basis in research training but, with many employers now expecting employees to be familiar with information management and to have an understanding of research findings, this is a useful combination whatever career path students may take.

Teaching, Learning and Assessment

Modules are drawn from those offered in masters programmes in the School of Social Science and from other Schools within the University, and students on this M.A. will be studying alongside students on those programmes. The pattern of teaching, learning and assessment will follow that of the modules chosen for the student's specific programme.

M.Sc. RESEARCH METHODS IN SOCIAL SCIENCE

Mode. Two years part-time, one year full-time. For core modules, specialist methods options.

The M.Sc. Research Methods in Social Science is a multi-disciplinary degree programme for those who want to acquire or extend a knowledge of research methods across a broad span of the social sciences. It may lead to the award of M.Sc., PG Diploma or PG Certificate. It also forms a strong basis for research students proceeding to an M.Phil. or Ph.D. (part-time or full-time).

Programme Structure

The core of the programme consists of five compulsory modules of research methodology. Key skills modules are also recommended. In addition students studying for the M.Sc. submit a dissertation of about 12,000 words and take one further taught module selected from the list of option courses available.

Core Modules

- Methodology and Research Design
- Qualitative and Quantative Research Methods I and II
- Research Context Process and ethics
- Advanced Research software and applications.

Options

- Depth interviewing
- The study of subjectivity
- Spatial Analysis for Social Science
- Action research and evaluation techniques
- Textual and visual analysis
- Social survey methods
- Epidemiology
- A range of options on statistical methods.

Entry Requirements

Applicants must normally hold an honours degree in the social sciences, or possess relevant professional qualifications.

Admission to the programme is by interview, based on the above criteria and on your understanding of and commitment to the aims and philosophy of the programme.

M.A. CRIMINOLOGY

Mode. Two years part-time; one year full-time.

Programme Structure

The part-time programme is taught over two years; the full-time programme is taught over one year. The M.A. is awarded to those students successfully completing all modules, including the submission of a dissertation.

Programme Content

- Issues in criminological theory
- White collar crime
- Feminist perspectives in criminology
- Research and evaluation in criminology
- Issues in criminal justice
- European perspectives on crime and the criminal justice system
- Crime and victimisation
- Prisons in context
- Racism, criminal justice and social exclusion
- Preparing for criminological research
- Dissertation

The programme provides students with an overall understanding of the nature of crime and its control, encompassing an in-depth investigation of the major theoretical paradigms of contemporary criminology; the methodological insights developed from each theoretical position; pattern and types of offending and criminal victimisation by age, gender, race and class and within given localities; the problems of research within criminology and the evaluation of crime control interventions; an analysis of the major institutions of the criminal justice system; the policy implications of each theory and the problems of implementation.

Teaching, Learning and Assessment

Knowledge from the programme is applied to the development of skills in applying particular theoretical perspectives to specific problem areas of crime and the institutions of crime control; developing an in-depth dissertation on a specific area of theory or research; designing research strategies into the understanding of crime and its control; evaluating crime control interventions. Modules on the programme are assessed by a 4,000 words essay submitted at the end of each semester. The dissertation, which is 15,000 words long, is submitted in September.

M.A. CRIMINOLOGY (Distance Learning)

Mode. Modular Study taught via internet. Programme usually completed in two years.

Programme Structure

There are two routes to gaining the M.A. Criminology through distance learning.

Route 1 : Students on this pathway take the two core modules (see below) plus three optional modules, and the module 'research in criminology' at the Middlesex University summer school. Finally students on this pathway will complete a 15,000 word dissertation.

Route 2 : Students on this pathway will not complete a dissertation and hence do not need to take the summer school module. Instead, students on this pathway will complete their M.A. Criminology by taking the two core modules plus six optional modules and write a project of their own choosing.

Programme Content

The following modules are available :

Core Modules

- Issues in Criminology Theory
- Issues in Criminal Justice.

Options

- History of Crime and Punishment
- Crime and Organisation
- White Collar Crime

- Organisational Context
- Youth Crime and Justice
- Probation Studies
- Indian Criminology
- Prisons in Context
- Dissertation.

The programme provides students with an overall understanding of the nature of crime and its control, encompassing an indepth investigation of the major theoretical paradigms of contemporary criminology; the methodological insights developed from each theoretical position; patterns and types of offending and criminal victimisation by age, gender, race and class and within given localities; the problems of research within criminology and evaluation of crime control interventions; analysis of the major institutions of the criminal justice system; the policy implications of each theory and the problems of implementation.

Teaching, Learning and Assessment

All teaching and learning is via the internet, use of specially written module guides and selected 'textbooks' are included in the cost of the programme. Each student is assigned a tutor for each module taken and a personal tutor who provides guidance throughout the whole programme. The assessment of all modules (except the dissertation and 'research in criminology') is by a 5,000 words essay submitted at the end of each semester. A 15,000 words dissertation is required.

M.A. SOCIAL POLICY

Mode. Two years part-time. One year full-time. Precise pattern of attendance depends on modules selected.

The M.A. Social Policy is a well established programme. It is of particular relevance to students currently employed in the public voluntary sectors who work in, or may wish to work in, middle management. It is also a good training programme for students who have a background knowledge and experience of social policies who wish to seek employment in the public or voluntary sectors or others who have a first degree in social science who wish to pursue advanced study in social policy.

Programme Structure

Students take four core modules over the two years plus two skills modules. Students may chose to do either a 15,000 words dissertation in which case they would take one option or a 10,000 words dissertation and two options. Alternatively, students may choose to complement the core and skills modules with four taught options.

When students are interviewed for the M.A., either the Programme Leader or the Admissions Tutor will discuss the composition and structure of their proposed degree with them. A Postgraduate Certificate and diploma are available as well as a Masters degree.

Programme Content

Core Modules

- The Development of Social Policy 1945-1979
- The Politics of Social Policy
- Public Sector Management
- Care in the Mixed Economy

Skills Modules

- Study skills
- Dissertation skills.

Options

- Disability; Key Perspectives
- Drugs : problems, policy and practice
- Racism, social policy and social justice
- Employment and education
- The voluntary sector
- Migration, theory and policy.

The options offered in any year depend upon student choice and staff availability.

Internship Modules

Work Experience Internships are available in place of options. The central concern of the programme is to reflect the many sided character of debates on the 'Welfare State' as we approach the 21st century. These debates relate to the respective roles of state, market, voluntary and informal provision of welfare, and the changing institutional context — at international, national and local levels, within which welfare needs are met. The focus on contemporary issues is combined with setting the changes in a broad historical context.

Teaching, Learning and Assessment

The programme provides an intellectual setting within which students can develop of fuller appreciation of substantive areas of social policy and improve their analytical skills. Students acquire the ability to analyse the formation and implementation of social policies and to use historical knowledge and evidence to evaluate contemporary discussions. Students will also gain the ability to present the results of policy analyses clearly and effectively. Dissertations offer an opportunity to undertake an in-depth study of a particular area. Teaching is via lectures, group discussion, case studies, workshops or seminars. Students are assessed by extended essays, examinations and, where chosen, a dissertation.

Entry Requirements

Admission to the programme is by interview, based on the above criteria and on your understanding of and commitment to the aims and philosophy of the programme.

M.Sc. PSYCHOLOGY AND HEALTH

Mode. Two years part-time.

The M.Sc. Psychology and Health equips students with a knowledge of theoretical issues in health psychology, increases their awareness of the complex multidisciplinary issues which this study entails, and enables them to carry out research in this area. It provides an account of the psychological, social and policy processes which relate to the promotion and maintenance of health and the prevention of ill health. The programme is innovation in placing emphasis on the psychological interface between the individual and society as a key issue in the determination of health.

Programme Structure

All students take two health psychology core modules, and a research methods module. In addition, students choose at least three further modules from a range of 9 options.

Programme Content

Core Modules

- Health Psychology Core courses A & B
- Research methods for health psychology.

Options

- Complementary medicine
- Health promotion
- Ageing and health
- Health policy
- Evolution and problems of modern medicine
- The psychology of addictive behaviours
- Health and the organisation of work
- Health, morality and medicine
- Introduction to counselling and therapy.

In the health psychology core courses a survey of research findings is provided together with an evaluation of current controversies. The programme is divided into the following areas; socio-cultural and psychological variables associated with good and bad health, models of health behaviour, stress and stress management, communication with health care professionals, illness representations, personality and illness, psychological aspects of disease prevention.

In research methodology it will be assumed that the students are already familiar with the basic principles of experimental design, descriptive and inferential statistics as normally taught on undergraduate psychology programmes; those with alternative educational backgrounds will be encouraged to remedy any gaps in their knowledge by private study, and tutorial assistance will be provided. The programme focuses on practical problems of research design in different areas of particular importance in health psychology which are not always covered in detail on undergraduate programmes (qualitative methods, questionnaire and survey methods, data analysis using computer packages). Students are also asked to give short presentations of their own dissertation plans.

Teaching, Learning and Assessment

The programme aims to increase students' knowledge of the historical and philosophical context of psychology as applied to health, of psychosocial aspects of health and of psychological mechanisms in health, illness and care; to increase awareness of ethical issues, values, policy issues and social issues such as gender, race, culture, and socio-economic factors; and to provide skills in designing, conducting, analysing and evaluating research on psychology and health. The programme utilises a variety of teaching methods appropriate to the diverse needs of its various units. These include lectures, seminars, student presentations, video presentations and practical work.

The M.Sc. dissertation is a central and crucial plank in the programme. It develops the knowledge and skills required to conduct independent research in psychology and health. Students select a research project, in consultation with the programme leader who allocates a supervisor. They may elect to conduct their practical work in the laboratory or in a relevant health setting.

M.A. PG DIP. SOCIAL WORK (DIP SW), M.A. SOCIAL WORK EMPLOYMENT ROUTE M.A. PG DIP. SOCIAL WORK

Mode. The M.A. PG Dip. Social Work Programmes offer students the opportunity to combine academic study with professional training. They lead to a university and (except the M.A. Social Work 'Top-Up') a Central Council for Education and Training in Social Work (CCETSW) qualification in social work (Dip. SW). They attract graduates already in social work posts who are seeking training as well as those making a change of career. They promote the importance of good social work practice and the development of appropriate and applicable knowledge, values and skills.

Programme Structure

The programmes are all clearly structured and prescribed. Most Social Work modules are year long, running for two semesters. All full-time postgraduate students follow the PG Dip. programme in the first year. On successful completion of all assessed work they can register for the M.A. programme in the second year. Placements (except the Employment Route programmes) run concurrently with academic study and are undertaken in a wide variety of practice agencies. Students on the Employment Route remain within their current place of employment (a contract between the student, the employer, and the university will be signed). Due to the nature of the professional requirements for the CCETSW award of Diploma in Social Work, students are required to attend all modules, except where Accreditation of Prior Learning as applicable to specific modules has been granted. APL for the M.A. Social Work 'Top Up' is available for candidates holding a compatible PG Dip. in Social Work (Dip. SW) from Middlesex University or any other approved UK University. All staff listed within the brochure as teaching in Social Work undertake supervision for M.A. students.

Programme Content

Core Modules

- Social work knowledge and skills
- Social problems and social policy
- Anti-discriminatory practice
- Social work law
- Applied psychology
- Community group and family work
- Social work research.

Pathways (Specialist Areas)

- Children, young people and families
- Adults with specific needs
- Working with offenders : criminal justice and youth justice
- Social work and palliative care.

Teaching, Learning and Assessment

Throughout the programme of study, students are encouraged, guided and expected to develop their own self-managed learning. They are expected to follow the programme's basic philosophy concerning the knowledge base of social work, its values and skills, while at the same time developing their powers of enquiry and critical appraisal.

Students are assessed by essays, projects, practice work and, for the masters programmes, a dissertation. Due to the special academic and professional aspects of these programmes, all modules and placements have to be successfully completed for the student to receive the award.

Accreditation of prior learning is available for suitably qualified applicants.

M.A. PG DIP. YOUTH JUSTICE, PROBATION AND APPLIED CRIMINOLOGY

Mode. The M.A. Youth Justice, Probation and Applied Criminology is designed to provide a thorough, critical appraisal of youth justice, probation and applied criminal justice. The programme is particularly relevant to practitioners working within the criminal justice system : youth justice, youth work, probation, victim support, community safety, crime prevention, drugs and social work.

Programme Structure

The part-time programme is taught over two years : the full-time programme is taught over one year. The M.A. is awarded to those students successfully completing a total of six modules, including the submission of a dissertation.

Programme Content

Core Modules

- Probation and applied criminology
- Youth justice
- Organisational context
- Research and applied criminology
- Dissertation module.

Teaching, Learning and Assessment

This masters programme provides, through taught inputs, self-managed learning and research, the theory and understanding necessary to develop specific knowledge about applied criminology in a variety of settings. Modules on the programme are assessed by a 4,000 words essay submitted at the end of each semester. It is also possible to undertake an internship in an appropriate cirminal justice agency to gain first hand experience, assessed by a portfolio. The final dissertation, which is 15,000 words long, is submitted in September following completion of the taught programme, including internship, if undertaken.

M.A. PSYCHOANALYSIS

Mode. The M.A. Psychoanalysis provides a postgraduate qualification for anyone with a clinical or an academic interest in the field of psychoanalysis. Accreditation is available for previous academic study and for prior clinical experience, and this enable certain students to omit the first year content of the programme; for example, someone who is already a trained analytical psychotherapist would have this option.

Programme Structure

The teaching programme is organised around a fortnightly case seminar, where practitioners from a wide range of orientations to psychoanalysis present their clinical work within the format of an academic seminar. A theoretical background to the seminars is provided by four semester long papers on the history of psychoanalysis, and on current issues in the analytical field. In addition a series of shorter papers on the relation of psychoanalysis to other fields supplements the case seminar series.

Programme Content

Core Modules

- The psychoanalytic tradition
- European psychoanalysis
- The relations of theory and technique
- Contemporary issues in psychoanalysis
- Case Seminars 1 and 2.

Options

- Psychoanalysis and feminism
- Psychoanalysis and problems of method
- Psychoanalysis and psychology
- Psychoanalysis and literature
- Psychoanalysis and its critics
- Psychoanalysis and philosophy
- Psychoanalysis and the social.

A further paper, Case Studies in Psychoanalysis is available, but only to people registered in the full-time mode.

In addition to the study of major figures of the psychoanalytic movement — Freud, Lacan, and Klein — the work of the second and third generation of psychoanalysts will be considered from the perspective of the central problem of the relation of clinical practice to analytical theory.

It follows from this primary orientation that current problems of the relation of psychoanalysis to philosophy, to psychology, to feminism, to sociology, and to literature are seen as consequences of the formulations given to questions in clinical analytical work. All these issues are addressed in the programme, which seeks to promote a wide-ranging consideration of the nature of psychoanalysis, as a pre-condition for the advancement of understanding and research in this field.

M.A./M.Sc. SUSTAINABLE ENVIRONMENTAL MANAGEMENT

Mode. The M.A. M.Sc. in Sustainable Environmental Management has developed from our previous long established programme in Conservation Policy. It is designed for those who wish to gain a high level qualification in the field of theory and application of sustainable management policy and who wish to explore further their interests in the links between global problems and local actions. The programme is likely to appeal to recent graduates from environmental disciplines and those already working in, or aspiring to work in, environmental non-governmental organisations, local authorities and environmental agencies and in the field of education; and overseas students with an interest in the global context of their specific environmental problems.

Programme Structure

The programme is two-thirds taught modules and one-third dissertation. Full-time students take three core modules in semester one, three option modules, including optional work experience, in semester two and complete their dissertation in the remaining four months of the year. Part-time students normally take two core modules in semester one, two option modules in semester two of their first year and the core research methods module in semester one of the second year. The sixth module may be taken in either semester one or two of the second year. Part-time students may take their programme over the longer period.

Transfer between full-time and part-time mode is also possible. The award of M.Sc. is normally available to those with a B.Sc. as their first degree. Requirements are at least one designated science option and the use or application of scientific methodology in the dissertation study.

Programme Content

Core Modules

- Environmental principles
- Environmental management policy
- Research and project methods
- Dissertation (Diploma or Masters).

Option Modules

- Project management
- Coastal-zone management
- Managing environmental conflicts
- Rural management
- Environmental Economics
- Environmental law (1 and 2)
- Tourism development and tourism impacts
- Work experience.

The programme aims to provide a holistic view of environmental management through the integration of social, economic and environmental perspectives. It begins by exploring key environmental principles based on interactions between society and the natural environment. At the same time, students study theoretical frameworks and their practical application for understanding contemporary environmental management policies. Case studies, including a residential field course, play an important part in studying the practical interpretation of sustainable management principles. Progress in implementing the Earth Summit agreements, including Local Agenda 21, is critically reviewed. The module in Research Methods equips students with a range of skills necessary to embark on research for their dissertations.

Teaching, Learning and Assessment

The programme provides a framework within which students can develop their specialist interests in the application of sustainability principles to various aspects of environmental management ranging from local community involvement in

Local Agenda 21 to global environmental issues. It aims to provide a sound grounding in policy analysis, as applied to a wide range of environmental policy areas, for those without a social science background. Students learn the ability to critically review policy implementation and the importance of monitoring and evaluation of evidence.

The programme includes a range of teaching methods in addition to lectures and seminars, such as workshops, group projects, student presentations and field work. Optional work experience, in place of a taught option module, enables recent graduates to develop, apply and extend relevant skills in environmental organisations.

Core modules are assessed by course work projects, reports and essays and some modules include components of verbal and visual presentation skills. The dissertation provides a focus for application of a range of approaches to the evaluation of sustainable management policy and practice.

M.A./M.Sc. SUSTAINABLE ENVIRONMENTAL MANAGEMENT

(Geography and Environmental Management)

Mode. The specialisation in Geography and Environmental Management is a very flexible programme which has been designed for those who wish to gain a higher level qualification in the field of Environmental Management with the opportunity either to select options from a wide range of topics, or to construct a more specifically focused programme which enables them to specialise in new areas or build on existing knowledge. The programme will appeal to recent graduates from Geography, Environmental Policy or Environmental Science subjects and those already working in a wide range of organisations, *e.g.*, local authorities, environmental agencies and in the field of education; and overseas students who wish to study principles and practice of environmental management as applied within the United Kingdom in order to relate them to their own country's needs.

Programme Structure

Masters students' programme is two-thirds taught modules and one-third dissertation.

Full-time students take two core modules and one option in

semester one; three option modules, which may include optional work experience, in semester two and complete their dissertation in the remaining four months of the year. Part-time students normally take one core and one option module in semester one, two option modules in semester two of their first year, and the core research methods module in semester one of the second year. The fourth option module may be taken in either semester of year two. Transfer between full and part-time mode is also possible. The award of M.Sc. is normally available to those with a B.Sc. as their first degree. Requirements are : at least one designated science option and the use or application of scientific methodology in the dissertation study.

Programme Content

Core Modules

- Environmental principles
- Research and project methods
- Dissertation (Diploma or Masters).

Option Modules

- Project management
- Environmental management policy
- Work experience
- Rural management
- Managing environmental conflicts
- Environmental economics
- Environmental law
- Tourism impacts and tourism development
- Environment and development in the third world.

Designated Science Options Include

- Coastal zone management
- Water quality management
- Hydroecology
- Environmental bio-monitoring.

The programme is anchored in the study of environmental principles, their historical context and their application and in research methods. Beyond this, students design an approved programme according to their interests and needs. Broadly based programmes could encompass different aspects of environmental management (*e.g.*, water management) combined with economic development or could focus on a particular theme (*e.g.*; Third World Development). The module in research methods equips students with a range of skills necessary to embark on research from a social or policy perspective for their dissertations. Specific scientific skills will be developed in appropriate option modules for M.Sc. students.

Teaching, Learning and Assessment

The programme provides the framework within which students can develop their specialist interests in the broad field of geography and its application to environmental management ranging from local to global environmental issues and management problems. It aims to provide a sound grounding in social research skills as applied to a wide range of areas for those without a social science background. The programme includes a range of teaching methods, in addition to lectures and seminars, such as workshops, group projects, student presentations and field work. Core modules are assessed by course work projects, reports and essays. Some modules include components of verbal and visual presentation skills. Optional work experience, in place of a taught option module, enables recent graduates to develop, apply and extend relevant skills in appropriate organisations.

The dissertation enables students to apply specific skills to a wide range of environmental management issues.

M.A./M.Sc. SUSTAINABLE ENVIRONMENTAL MANAGEMENT (*Water Management*)

Mode. The specialisation in Sustainable Water Management builds on central themes of water science and management. It covers a range of disciplinary perspectives in order to develop a balanced appreciation of the complexity of integrated water management, which builds jointly on environmental science and policy studies. It has been developed to provide students with a sound understanding of the principles and practice of both water processes and the social, economic and political factors involved in the management of water in order to enable them to adopt an integrated approach. The purpose of the progamme is to provide the students with knowledge and skills required to obtain employment in the water industry or within relevant local or Central Government departments.

The programme is designed for graduate and practitioners with an interest in water management including river corridor management, and integrated coastal management; those seeking employment within the water industry, in private consultancies or with central or local government regulatory agencies; professionals already working within the above agencies, or individuals who are actively involved in some form of environmental/river management who wish to increase their conceptual awareness and skills levels; graduates from overseas who wish to study water management in the UK and Europe in order to place this experience within the context of their own country's water management problems.

Programme Structure

Masters students' programme is two-thirds taught modules and one-third dissertation.

Full-time students take three core modules in semester one, and one in semester two and two option modules in semester two and complete their dissertation in the remaining four months of the year. Part-time students normally take two core modules in semester one, one core and one option module in semester two of their first year, and the core research methods module in semester one of their second year. The sixth option module may be taken in semester one or two of year two. Transfer between full and part-time modes is also possible.

Programme Content

Core Modules

- Research and project methods
- Environmental management policy
- Water quality management
- Dissertation (Diploma or Masters).

Designated science options include

- Coastal zone management
- Hydroecology
- Water and waste-water treatment
- Physico-chemistry and modelling of water quality.

Option Modules

- Environmental economics
- Environmental conflict resolution
- Environmental law.

The programme begins with an introduction to environmental management and policy generation processes. This is complimented by an in-depth study of water quality planning and management systems. At the same time, those following a full-time programme will receive an introduction to research methods to allow the initiation of ideas and research on an agreed dissertation topic.

The specialised modules which follow provide an in-depth look at water quality management; public participation and liaison processes between public, industry, the water agencies and the regulator; river corridor and coastal zone management; the biological quality of rivers and streams; and water quality modelling. These modules focus on issues associated with the use of river corridors, flood plains, flood management and low flows, ecology, biology, conservation and water pollution control.

Teaching, Learning and Assessemnt

The programme provides a focused framework within which students can apply the principles of integrated water planning to their specialised areas of interest. It aims to provide a sound grounding in the science and policy of water planning. The programme includes a range of teaching methods in addition to lectures and seminars, *e.g.*, laboratory work, field visits, visits to water agencies and visiting speakers who are current practitioners.

Core modules are assessed by a mixture of course work and examinations.

M.A. SUSTAINABLE RURAL DEVELOPMENT

Mode. The M.A. in Sustainable Rural Development provides students with the necessary theoretical and conceptual frameworks to properly interpret the process of social change in rural areas, and gain sufficient knowledge of rural activity to evaluate, formulate and manage the implementation of rural development programmes. The programme will appeal to : graduates and practitioners from developing countries who wish to develop their understanding and skills in managing rural development programmes in their own countries; recent graduates in geography, economics, development studies and area studies who wish to develop a deeper understanding of development policies in third world settings; those working for, or aspiring to work for, development organisations and voluntary agencies who wish to develop their understanding of rural development processes and who wish to equip themselves further; those who are working in private sector consultancies and non-governmental organisations and who wish to enhance their expertise and, educationalists and researchers who wish to develop their understanding of rural development processes and skills.

Programme Strucutre

Masters students' programme is two-thirds taught and modules and one-third dissertation.

Full-time students take three core modules in semester one, two core and one option modules in semester two and complete their dissertation in there remaining four months of the year. Part-time students normally take two core modules in semester one and two in their first year, and the core research methods module in semester one of their second year. The sixth option module may be taken in semester one or two of the second year. Work experience may be accredited in place of an option module. Transfer between full and part-time modes is also possible.

Programme Content

Core Modules

- Research and project methods
- Project management
- Rural policy and planning studies
- Development economics/studies
- Environmental management policy
- Dissertation (Diploma or Masters).

Options Include

- Environment and development in the Third World
- Culture and development in the Middle East
- Culture and development in Africa
- Spatial analysis/advanced spatial analysis and GIS
- Water quality management
- Gender, society and development in the Caribbean.

The programme seeks to develop an integrated, interdisciplinary approach combining theoretical and scientific skills with practical case studies and field work. The programme considers the various resource components; environmental, social and economic and how they can be most effectively managed to achieve the most sustainable rural systems. The purpose is to provide a sufficiently broad area of specialisation for entry into employment in policy formulation, implementation and administration or as researchers and teachers.

The programme is based upon a number of core areas including development economics, rural social development and research methods and project management.

Teaching, Learning and Assessment

Teaching adopts an integrated interdisciplinary approach combining theoretical and practical skills with case studies. The programme considers the various resource components; environmental, social and economics indepth and how they can be most effectively managed to achieve sustainable systems.

M.Sc. IN ARBORICULTURE AND COMMUNITY FOREST MANAGEMENT

Mode. The Masters in Arboriculture and Community Forest Management offers several different and flexible study routes to enable people from Arboriculture, Countryside Management and related professions to gain higher level qualifications.

Applicants can select from either a largely taught programme through to an entirely work based programme, with the opportunity to gain an Advanced Diploma as part of the qualification.

The focus of the programme is to develop the understanding and application of the management of staff, the environment and resources within the student's professional field.

Programme Content

The essence of the programme is the understanding and application of management skills to a variety of arboricultural and countryside management situations. It will enable participants to understand and apply issues of sustainability to their work, and to assess the environmental impact of potential operations. The programme will also assist participants to demonstrate the effective management of resources and staff and show their understanding of the inter-relationship between these different management strategies.

Through adopting a work based learning approach, participants will be able to claim credit points for prior learning, and agree credit for their current demonstration of learning and application of new skills in their work; and identify and carry out an agreed research dissertation which demonstrates the development of new ideas and approaches to their profession and within their workplace.

Students following the work based learning approach will be assigned an academic tutor who will meet, advise, discuss and assess their work, either at their workplace or at the university.

Study Pattern

Full-time taught programme

Following award of credit for prior learning, participants will study the Research and Project Methods module — as well as other modules selected from : Environmental Principles, Project Management, Environmental Management Policy, Environmental Economics, Water Quality Management, Rural Social Development, and others from the University's common academic framework, before carrying out a research dissertation.

Part-time Work based programme

Participants will agree with their academic tutor the modular content of the programme, selecting from the following :

Advanced Diploma

A series of short intensive taught sessions followed by work based projects addressing topics such as Environmental Impacts, Legislation Ecology, and Financial Management, Marketing Business and Finance.

Prior Work Experience

This module will allow participants to compile evidence to demonstrate their existing learning and gain credit.

Work Experience

Based on the outcomes of the prior work experience module or the Advanced Diploma, participants will agree with their academic tutor the nature of the work skills and the understanding needs to fulfil the requirements of the programme.

Research and Project Methods

This is a compulsory module and will prepare students for the research dissertation. It may be studied as a taught programme or as a resource based module.

Project or Research Dissertation

This is the final module of the programme and requires the participant to devleop a discrete research project closely allied to their profession and/or their work. It should be innovative, detailed, demonstrating insight and comprehension.

Assessment

Assessment is through extended essays, portfolios, presentations, and a dissertation of approximately 15,000 words.

Attendance

The taught programme is run from Middlesex University mostly at the Enfield site. The Work Based Learning modules will require meetings between the participant and their academic tutor at an agreed location. This study route can be started at any time of year.

OTHER PROGRAMMES PG DIP. M.A./M.Sc. INTERPROFESSIONAL HEALTH CARE

Mode. This part-time programme is designed to meet the rapidly changing needs of health and welfare practitioners as we enter the 21st century. There is a growing recognition of the need for practitioners to be both reflective, and responsive in dynamic and innovative ways to changing demands. Interprofessional understanding and collaboration is a key feature of current and future service provision. This programme enables practitioners from a wide range of professional backgrounds to work, learn and develop new skills in a friendly and progressive environment.

Programme Structure

This part-time programme is structured around four core modules, two option modules and a triple weight dissertation. In year one two core modules are taken, and in year two one core module and a triple weight dissertation.

In addition to these core modules an independent learning module is taken during the taught part of the programme, prior to the dissertation. Students are required to undertake two option modules during the programme, and may access modules from the Interprofessional Health menu, or from outside of this menu in negotiation with the pathway coordinator. These options may be taken in either the first or second year.

Programme Content

Core Modules

- Values, Ideas and Beliefs in Professional Practice
- Methods of Critical Enquiry in Health Care Sciences
- Health, Welfare and Citizenship
- Interprofessional Development and Progress.

Option Modules

Any of the Interprofessional modules, at the appropriate level, may be assessed. Modules from outside the subject area may be assessed with the agreement of the pathway coordinator.

Teaching, Learning and Assessment

This masters level programme equips students with the knowledge, skills and competencies to work effectively within the interprofessional team in the field of health and welfare. A range of teaching/learning strategies are utilised including lectures, workshops, seminars, role play, case studies and tutorials. Students are assessed through essays, projects, case studies, examinations (according to modules chosen) and a research dissertation.

Entry Requirements

Applicants should hold a first degree and/or a recordable professional qualification. Non-graduates will be expected to produce evidence of the ability to study at postgraduate level.

M.Sc./PG DIP. MENTAL HEALTH INTERVENTIONS

Mode. The programme will equip mental health professionals with the theoretical research knowledge, core clinical skills and values necessary in the care and treatment of people with a variety of serious mental health problems. On completion of the programme, graduates will be capable and advanced contemporary clinical practice characterised by the application of research findings.

Programme Structure

This part-time programme extends over a two year period. The M.Sc. is awarded to those completing all modules including the submission of a dissertation.

Programme Content

- Contemporary Developments in Mental Health
- Research Methods in Mental Health
- Case Management
- Service Development
- Psychosocial Interventions
- Medication Management
- Dissertation Supervision Workshops.

Entry Requirements

Applicants should hold a professional qualification in nursing, social work of allied discipline and either hold a good honours degree in a relevant social science subject or have a minimum of 2 years professional experience in the mental health field. (Those not holding a first degree should have undertaken a level three module in research methods or equivalent.) Non mental health professionals with good, relevant honours degree and at least three years' experience in Mental Health practice are also eligible to apply.

Applicants must be working or have access to clinical practice as they will be expected to undertake substantial amounts of clinical practice during their periods of attendance on the programme.

M.A. LOCAL ECONOMIC DEVELOPMENT

Mode. The M.A. Local Economic Development provides a thorough grounding in the interpretation and analysis of processes, issues and policies relating to local economic development. Recent economic and political change has increased the priority given to a wide range of local agencies pursuing policies promoting economic development and employment creation. This programme provides a unique opportunity for those wishing to gain a high level qualification in the specialist policy field. It will be of interest to all those engaged in promoting economic development and employment at the local and regional scale as well as those interested in obtaining a deeper understanding of policy debates and issues with a view to developing a career in this field. This includes members of local authorities. TECs, Central Government departments, regional development agencies, voluntary agencies, planning consultants, and recent graduates.

Programme Structure

Students may study on a full-time (one year) or a part-time (two year) BASIS. Those who do not wish to study for a masters degree (180 credits) may study for a certificate (60 credits) or a diploma (120 credits). For M.A. students studying part-time, attendance is generally required on one day per week (currently from 2.00 p.m. 7.00 p.m. on a Tuesday), across three and a half 12 week semesters.

Programme Content

The programme consists of three types of modules : compulsory, specialist option and elective :

Compulsory Modules

These modules provide the core curriculum to the Programme. Forces Shaping the Local Economy identifies the main economic processes and factors acting upon local economies whilst key debates in Local Economic Development provides an overview of the history, institutions and major strategies of local economic development. Students are also required to complete Analysing the Local Economy, which introduces methods of local economic analysis, and a Masters Dissertation on an approved topic relating to local economic development.

Specialist Option Modules

These provide opportunities for shorter in-depth study of particular aspects of local economic policy and include : European Local Economic Policy, People Based Employment Inward Investment and Local Economic Development, Technology and Local Economic Development, Labour Market Policy, Small Firms and Local Economic Development.

Elective Modules

These consist of wide variety of other complementary modules available within Middlesex University and include areas such as research methods, computing or management skills.

Teaching, Learning and Assessment

Teaching methods include lectures, seminars, workshops, tutorials and day visits. Student participation is actively encouraged at all times. The teaching staff are experienced researchers in the field and are complemented by visiting speakers. All modules are assessed by course work which includes essays, reports, projects, seminar presentations and a dissertation.

OPPORTUNITIES FOR RESEARCH

The Institute of Social Science Research covers a wide variety of research activities of national and international standing. Research centres are the focus of much advanced-level work. These consist of high profile teams of academics, research fellows, support staff and research students. Often they adopt an applied interdisciplinary orientation, and produce output in terms of peer reviewed publications, policy advice, generic procedures for decision support, and critical independent reviews and analyses of social issues. Researchers are active in several areas — psychoanalysis, gender issues, health, racial equality, criminology, social policy, environment and business, hazards and sustainability.

Research degree completion requires motivation and discipline. But it is also a creative exercise : exciting, satisfying and for most students very challenging. At Ph.D. level in particular, the research degree process offers many opportunities for conference participation and exchange of ideas. Titles of Ph.Ds. recently awarded include Late Quaternary Alluvial Fan, Debris Cone and River Terrace Development in the MacGillycuddy's Reeks, South West Ireland; Poverty, Community and Health : Social Networks as Mediators between Poverty and Well Being; The Car and Crime : Critical Perspectives; Telling my time and Finding My Place; and Rhetoric or Reality? A Critical Investigation of the Market Model for Community Care.

Study Patterns

Research degrees may be taken either full or part-time in any subject in which the staff specialise. Full-time students would typically take at least three years in the case of a Ph.D. Part-time students would generally take nearly twice as long. Most of the student's time is spent developing, carrying out and writing up a research project. Projects may vary from research involving extensive field work or social surveys, to those based on pre-existing data, carried out largely by computer analysis, to those of a more theoretical nature. The aim is to produce a written thesis, which is a sustained argument based on a thorough review of evidence and critical enquiry offering original insights at a theoretical or empirical level. Once the thesis has been submitted, there is a two stage examination : the thesis itself is examined by internal and external examiners who also discuss the thesis with the candidate in a viva.

During the first year of M.Phil. and Ph.D. studies, students are required to undertake taught courses in research methods drawn from the M.Sc. Research Methods in Social Science. Other courses from across the university may be taken if they are relevant to students' research. Ph.D. students register for an M.Phil. and transfer to a Ph.D. later — subject to satisfactory progress.

10

CENTRE FOR HUMAN ECOLOGY IN ACTION

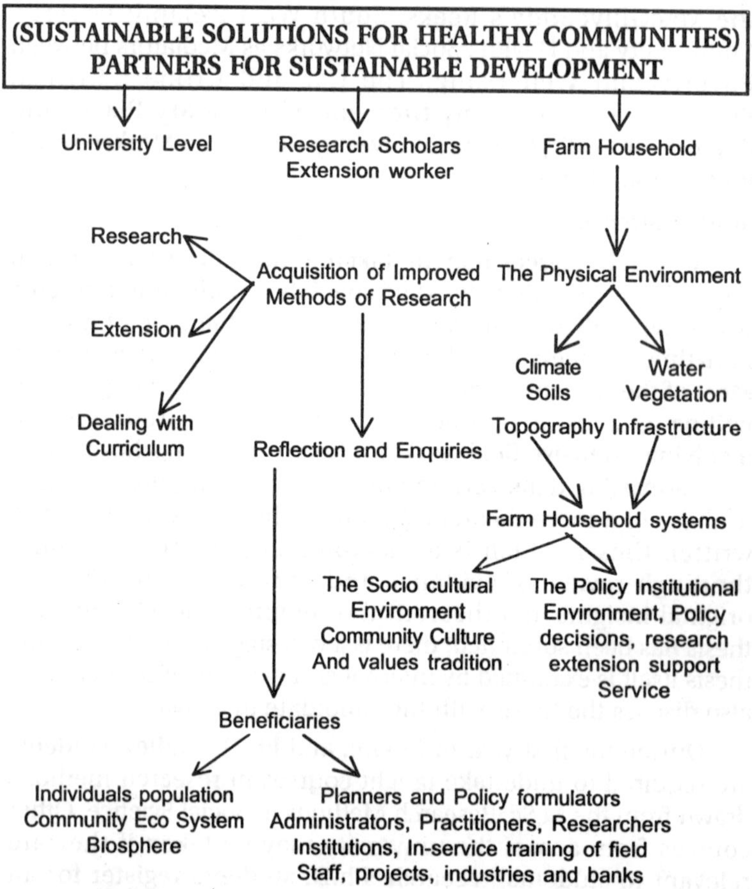

University level approach for reaching the un-reached through research, extension and education for training programmes for health and Human service.

RESEARCH METHODOLOGY FOR STUDIES ON WOMEN AGRICULTURE AND RURAL EXTENSION THROUGH EDUCATION AND TRAINING PROGRAMMES WITH SPECIAL REFERENCE TO PRE-SERVICE GRADUATES ON HOME SCIENCE

(Linking Home Science Curriculum with Farm Women and Community)

- Reorganisation of Home economics to human ecology on Policy framework to reform
- For the cause of Education system for Future.

Introduction

Historically, a major focus has been on the micro environment — home and society. As society technology, and home science institutions have evolved, increasing attention has been given to other settings in which needs are met and to interdependence within and across family, home, work, business and institutional environment. The history provides a sense of the holistic and ecological nature of the home science colleges and shows the ways in which inter-related functions and components are connected.

How the concept and philosophy of human ecology can be implemented through teaching research and outreach? From its inception human ecology has focused human being relationship to the material goods and social and economic system necessary for living. Over the courses of its history, human ecology will seek and apply knowledge from many sources and disciplines helpful to individual and family in meeting everyday needs and addressing significant human environment problems. The focus has been and continues to be on the activities of everyday life associated with meeting recurring sustenance needs for food, clothing and shelter and human development and relationships within various environments such as the family work place market systems, and other community settings.

The educational institutions have clear goals that are compatible with national policies and defined channels which users can make their demands known, it is, still important to recognise that the tasks of producing and transferring relevant technology are carried out by two groups of people —

researchers and technology transfer workers — and that these groups differ in attitude, interests and training.

There are series of tools and mechanisms which management can use to promote links among researchers and between researchers and technology transfer workers. In my view, committees, liaison persons and coordinating units often have a difficult time creating effective links, unless they have power and authority or control a budget.

We have to be very precise about what kind of links are needed, at what level they should be established, and what goals we hope to achieve. When we need so many linkage mechanisms that an organisation becomes excessively complex, we have to start thinking about reorganisation rather than creating more position or units in the organisation.

The goal of developing links to build bridges between researchers and extension workers should not make us forget that specialisation and division of labour have contributed to growth throughout history. Sometimes a bit of competition between actors will lead to improved performance by each of them, as long as the social institutional goals are clear for everyone.

In Brief

Several changes have taken place in the organisation and functional set up of Institute of Home Economics and in their research and development programmes. Commensurate in these, there has been no proper, emphasis towards development of management skills of faculty members to make national research programmes more efficient and effective Projects for R&D activities are time bound, and lay high priority to meet challenges such as poverty alleviation, food and nutrition security, natural resources conservation and overall sustainable development of agriculture and human resource; efficient managers are required. Thus need was felt to organise a Second Congress on Human Ecology I am quite hopeful that this Congress will provide good opportunity for the faculty of interdisciplinary and applied sciences to address the various issues and challenges faced and impact evaluation and assessment. There is also and also the need to link research to organisational and financial management; with planning HRD needs, sharing management and impact evaluation and assessment.

I hope that in these deliberations of this Congress many of the above mentioned issues will be discussed and crystallised and recommended for the follow up action. This will meet the requirement of linking research with involvement in research programme planning/governance at different stakeholders including private sectors and NGO's and farmers.

History

- Development of Home Science Education influenced by English System. 1950 Home Science Status brought to the university level. American universities were used as model. Spread to 96 institutions in India.
- UGC is not directly concerned with the setting up of Home Science Institutions.
- Planning Commission in 1950-51, followed by series of 5 years plan with the objective of planning :
  - — To increase production, expansion of employment
  - — To reduce inequality of income and wealth
  - — Major element free and equal society and their relevance to Home Science to increase people's capacity to work.

Methodology : Extension education and community organisation.

Extension : Techniques for purposeful action, to gradually bring about a gained technological changes.

Overview

In social sciences methodology is taken to be a discipline, bordering on philosophy, whose function is to recommend and examine the methods which should be used to produce valid knowledge. Methodology lays down procedures which are justified or criticised by means of philosophical arguments.

The project aim is to go beyond, mere descriptions of various methods, to examine the theory, logic and philosophy of the methods and to explore their implications. The questions of methodology proper, are linked both with the theoretical issues related to philosophy and to applied issues associated with various theoretical positions like positivism, phenomenology, structuralism, post-structuralism and so on.

However, there are critiques of objective research and positivist methodology in Social Sciences. The widespread acceptance of conventional methodology has been seriously questioned both on theoretical and political grounds by some Indian scholars.

In the past 15 years, developing countries' concern about persistent rural poverty and malnutrition. Unsustainable agricultural production, lack of food self-sufficiency, and swelling urban populations have become increasingly acute. In response, many governments have undertaken major initiatives to help resource-poor farmer households improve their living standards and productive capacity. These farmers make up not the majority of producers, but also a major share of the world's poorest and most vulnerable people.

The challenges for meeting the needs of resource poor farmers is to develop a greater capacity to facilitate effective interaction between researchers, technology transfer workers and resource poor farmers. This requires :

- Shifts in research policies and priorities
- Changes in the organisation and management of research and technology transfer agencies
- The development of strong links between these agencies and resource poor farmer.

Strategy

Development of strategy for food production, income generation through generic biodiversity, natural resource management and education to improve the environment of the community on sustainable basis.

Project planning and project preparation are vital. Efforts to ensure this need, are to be sustained. Following obstacles are faced in course of their improving the standard of living through new plans and policies are :

Improper training, insufficiency of technical staff and supervising facilities, lack of need based, non-availability of suitable mechanism to determine the appropriateness of the technology; insufficiency of government machinery to provide basic needs like supplying the improved techniques according to the demand on time, non-involvement of local people and organisation, shortage of funds, need of proper S&T policy measure; non-availability of technical advise, rectifying and replacing the equipment in use, lack of political will. Therefore, accurate effort needs to be implemented for the success of policy and programmes.

To linkages for tools and mechanism :

- Appropriate policy implementation from grass-root level
- Training facilities for research, extension and outreach
- Congress, Symposium
- For interaction.

Objectives

Institutionalisation on sustainable basis — The training programme on-farm, off-farm and non-farm sectors for the empowerment of tribal women in particular and human resource development as a whole.

To formulate the Home Science curriculum for conducting studies on women agriculture rural extension through education and training programmes with special reference to pre-service graduates :

— to fight poverty with passion and professionalism for lasting results.

— to help people, help themselves and their environment by providing resources, sharing knowledge, building capacity and foreign partnerships in public and private sectors.

— to be an excellent institution able to attract, excite and nurture diverse and committed staff with exceptional skills who know how to listen and learn.

Purpose

To assist University Departments/Research Institute/ Professional bodies/Individual Scholars for conducting (i) Research including action, research, (ii) Workshop/Seminars, and (iii) Publication of Studies and documentation.

Science and Technology can be utilised to upgrade

traditional skills, introduce new opportunities of work, reduce drudgeries faced during various household chores, improve the environmental and living conditions, reduce health problem. In fact, Science and Technology is capable enough of solving many problems of the people, if applied amidst their own environmental and cultural contact without creating any social and economic imbalances.

Environmental protection and development are the two facts of the same major issue. Environment and development are interdependent and cannot be separated. The questions related to environment are closely linked to human economic and social activities. The development of mankind cannot be sustained if problems of resources and environment are neglected. Likewise, protection of environment in isolation from economic development and technological progress has no basis. Therefore, we must trace back to the root cause of environmental degradation and find the best way to protect environment integrating it into economic and social development.

To assist those reputed social science research agencies which have a proven expertise and willing to undertake purposeful studies on the economic development, problems and requirements of Scheduled Castes and also on the impact of the implementation of various economic programmes undertaken by the governmental bodies through instruments of economic development.

To assist and develop linkages with governmental organisation for undertaking research including evaluation studies/projects, workshops and seminars helpful in orienting developmental programmes.

PROJECT : STUDIES ON WOMEN AGRICULTURE RURAL EXTENSION THROUGH EDUCATION AND TRAINING PROGRAMMES WITH SPECIAL REFERENCE TO PRE-SERVICE GRADUATES

Introduction

Many development programmes in the field of agriculture and rural extension have been launched by Government of India, non government organisations, education and training institute, professional association, scientific bodies, community organisations and also whole range of official agencies by the participation in the campaign or in partnership. The training and education may be able to build a network of researchers spread not only throughout India but also in the Asian countries as well. Research on education and training programmes will take lead and stay in forefront of research programme. In turn; this would establish an intellectual climate in the faculty of interdisciplinary applied sciences which would encourage the faculty to forge collaborations amongst themselves and with other scholars and practitioners in this effort to create a new knowledge base.

Finally a distinct expectation is that if the faculty is to participate in such a major creative intellectual process, it will have a powerful impact on the quality and depth of our educational and training resources besides of course, our research capacity.

Long-term Goal

Human resource development with special reference to in-service and preservice graduates in the light of women Agriculture and rural extension to achieve the following aims and objectives :

Aims and Objectives

Impart relevant education and training to young men and women for managing income generating and development activities on behalf of rural producers;

Offer training courses for policy makers, directors, general managers, and those in charge of specific managerial functions in rural enterprises and projects;

Conducted research on operating problems in the rural sector in order to help improve the management of rural enterprises and projects; and

Undertake basic research into the process of rural urban management to augment the existing body of knowledge on the subject.

Technical Programme

Duration — 3 Years

Pre Feasibility Study on Curriculum was conducted for

(a) Improvement in the existing programme and development of new programmes (in collaboration with Larenstein International Agriculture College, The Netherlands,) which is as follows :

  (i) Two programmes can be implemented

    (a) Pre-service training and — Graduates, B.A., B.Sc., B.Com.

    (b) In-service training to give boost to the rural development. — Already employed in rural development sector, NGO etc.

(b) Implementation of programme — 1 year Postgraduate Diploma

- Review of the contents and its constraints
- Feasibility of curriculum, modification if possible
- Development of curriculum and approval of academic council/research committee of the University.
- Study on inverter resources, to obtain technical support
- Organise work, determine state of art of previous curriculum if any including history etc.
- Recruitment of curriculum developers.

(A) Teaching and research activities. One year (In-service, Pre-service) programme for development practitioners aims to providing opportunities for young officers in rural development and producers' organisations to strengthen their professional management skills.

Three months General Management Programme provides opportunities to middle and senior level managers in rural producers' co-operatives for mid-career reorientation and training.

The six week Management Appreciation Programme for Voluntary Agencies provided exposure to the management. Short duration Management Development Programmes, Seminars and Workshops bring a large number of development professionals to the department every year for specialised training in the field of human resource management in the field of Human ecology.

(B) **Implementation** as per the modules (Modules attached)

Target group — NGO, Capart Projects

Phase-II

(C) Impact analysis — validation and implementation (Pilot testing)

- Conduction of the internal evaluation of the written curriculum with the fine tooth comb.
- Produce curriculum prototype.
- Testing of curriculum in terms of aims, rationales, objectives performance criteria, grading, contest entry characteristics, instructions, management of diversity, programme evaluation and production.
- Implementation, field testing in terms of evaluation, effectiveness acceptability, efficiency, motivation and vulnerability.
- Replication in other states within country, neighbouring countries like Bangladesh, Sri Lanka, Pakistan etc.

(D) Popularisation

- Implementation of programmes in Intra and Interstate, traditional and agriculture universities and colleges of the country and neighbouring countries.

Description of the Programme

The programme of the course has a theoretical and a practical component. The experiences of the participants is the point of departure. The course is divided into a number of modules during which the participant is enriched with new theoretical insights and practical skills. In each of these modules rural development, extension education contribute to the specific theme being studied. In this way an integrated unity is achieved. The modules are result oriented (practical implications and applications are an integral part of the modules). Apart from this, core programme is of special interest for the participants and can be attended through the position papers and the special project.

Design

All the programmes have been designed to acquire a strong understanding of the perspectives, concepts, tools and techniques in the respective disciplines and exposure to fundamental principles, framework and concepts in different functional areas and management in the field of Human Ecology.

COURSE CONTENTS (1 YEAR DIPLOMA COURSE)

Module I (Duration 3 Months)

Total Marks : 250
(200 Theory & 50 Practical)

1. *Introduction* to Agriculture, Development and Environment. (50 marks)

   Role of agriculture and Ecological related issues.

   Introduction, Tropical climate, Rainfall, Vegetation Soil-Physical properties, Ecological factor Vegetation-soil flora and fauna, Conversion of tropical rain forests, tillage farming system towards improvement in tropical agriculture, Science development and environmental problems, Role of ecological destruction in emerging patterns of poverty, Environmental problems in hills *e.g.*, Garhwal Environmental pollution of pesticides and associated health hazards, Environmental education, environmental planning, environmental toxicology.

2. *Nutrition,* hygiene and related issues like health and well-being. (50 marks)

   Nutritive value of Indian foods, Dietary Principles, Energy requirement, Science of the food, Applied nutrition — Nutritional consideration of groups of people — infant, school going children, adult — women pregnant and lactating, old people, Vegetarians and vegans, Meal planning, Nutritional problems. Introduction to Microbiology-Pathogens, spoilage and inert organism, control and growth of microorganism, food poisoning — chemical, biological and bacterial and its preventions.

   Hygiene in kitchen, waste disposal, pest control, hygienic handling of food, storage of food, Diseases — food and water borne, Food spoilage and prevention.

3. *Women and Development* — A psycho-social perspective

with special emphasis on rural developments and its determinants. (50 marks)

Socio-economic status of women problems and perspectives. Women and social structure, Organisational role (programmes in the state) in uplifting the status of women and its legal political dimensions.

Economic development and politicisation role differentiation of female agricultural workers, women and family, and income generation activities of women with case studies.

Women crisis and constraints : Economic security of women — employment quality, Women and social structure, Status of women and socio-economic programmes under the 5-year plan, Problems of working women, Women's development and the ICDS, Women's legislative measures, Role of voluntary organisation in uplifting the status of women economic discipline and the labour market. Women's productivity in agriculture system small scale enterprises and women, Case studies for income generating activities.

4. *Knowledge* of Basic computer. (50 marks)

   DBase

   Word Processing

   Application programme

5. *Practicals* (50 marks)

   Questionnaire development for rural setting with projection of any of the topics of the module I.

Module II (Duration 3 Months)

Total Marks : 250

(200 Theory & 50 Practical)

Extension Education and Communication Techniques

1. *Extension Education* (50 marks)

   Importance of Agriculture Extension in the agriculture development process. Extension education systems, community development, social and cultural factors in extension, social structure, social division, age, sex and religion.

2. *Rural Sociology* (50 marks)

   Village pattern — mode of living, social welfare, social organisation, group mobilisation and leadership, problems of caste and creed. Management of farmers organisation — Local Panchayat Raj, Mahila Mandal, Youth club, co-operatives, non government and government organisations leading to Community development.

   Extension teaching methods — Selection of methods; classification of methods — organising group discussion. Farm and home visit. Method demonstration. Result demonstration. Newspaper, Leaflets and pamphlets, Circular letters, Blackboards, Photographs, flash cards, Flannel graphs, Films slides, Specimens, Models and exhibits, Radio talks, Slide projector, Opaque projector, Overhead projector, Movie projector, Tape recorder, Camera. Relative effectiveness of methods.

3. *Communication* and Audio Visual for mass media development, Communication concept and channels, common economic interest groups, formal and informal leaders, social expectations, farming systems, land tenure, inheritance, ceremonies and festivals, social and cultural change, population growth, economic factors, social and cultural factors in agricultural change, relationship between communicator and communicatee, feedback and communication, through written words and other teaching aids. Techniques for data collection. (50 marks)

4. *Management* and extension (50 marks)

   Programme planning and intervention programmes for women and children in rural, urban and tribal areas with special reference to agriculture and total well-being.

   Designing extension programmes for rural development. Extension's role in adapting and evaluating new technology for farmers, introducing technological change; the social setting.

   Behaviour Research designs and variables, sampling techniques, reliability and validity.

6. *Practical* (50 marks)
   Development of Questionnaire, Pamphlet, Bulletins and use of mass media.

Module III (Duration 3 Months)

Total Marks : 250
(200 Theory & 50 Practical)

Farming systems & Non farm activities — Analysis of the problems, priorities and interventions.

1. *Farming Systems Development* (50 marks)
   Introduction, The concept and approach to farming to farming system development, Preparatory Arrangements when starting FSD, Farming system zoning, Exploratory Diagnosis and Analysis, Verification, Identifying Constraints and Potentials, Evaluation of Potential improvements, on-farm testing, Planning interventions and implementing PSD.
2. *Production and Operational management* (50 marks)
   Vegetable and fruit crops, Mushroom cultivation, Post harvest (Storage and Processing), Poultry keeping, Apiculture.
3. *Managerial Economics and Financial management* (50 marks)

   Concepts of demand and supply functions, elasticities, demand forecasting, pricing decisions market structure analysis, business risk, consumer behaviour etc. Money management including saving schemes and investments.
   Techniques and tools involved in the analysis of financial statements, management of working capital, investment analysis and financial controls.
4. *Management oriented Farming* (50 marks)
   Management and special emphasis on cost benefit analysis of the projects, demand for female labour, child care, management for informal and formal enterprises. Impact of different options for market-oriented production. Participants study the case of a market oriented demand.

5. *Practical* (50 marks)
   Farming system study in the peripheral village of Delhi.

Module IV (Duration 3 Months)

Legal Environment and Optionals

Total Marks : 250
(200 Project work and 50 Theory)

Legal Environment (50 marks)
(35 Theory and 15 Practical)

Law and society (law relating to women and children especially women and child labour), knowledge of the IPC, the Evidence Act, CPC, CrPC, the Commercial Laws, Torts, Motor Vehicles Act. Consumer Law, Environment Laws, Regulatory Laws, Organisational Laws, Land and Water Acts, PFA, Essential Commodities Acts, Taxation Laws, Regulatory Acts, Property laws, Inheritance law etc.

Optionals for further training in income generation
(200 marks)

Biotechnology. Tissue culture techniques, spore production (spawn/yeast and organic farming) use of microorganisms *i.e.* Production of bio fertilizer and single cell protein culture, preservation.

Other farm activities — Mushroom farming, floriculture, apiculture, poultry keeping and non conventional energy sources, vegetables and fruit production, dairy management and post harvest processing.

Economic Botany — Medicinal and Economic plants — Religion.

Non-farm activities — Tiny enterprises, Upgrading of traditional textile, handicrafts *e.g.* dyeing and printing, embroidery, applique, weaving, garment construction etc.

Creche, Non formal preschool education.

Economically productive projects for family and community — food based, textiles, handicraft based etc.

11

Centre for Social Justice and Empowerment

(For New Data and Charting a Course of Change)

Foundation Mission

"To help people, help themselves through the practical application of knowledge and resource to improve their quality of life and that of future generation."

Key Issues

- Why do the people migrate?
- Why do people voluntarily emigrate from the village, city and country?
- Why people are forced to emigrate from country?

HOW DO THE INVESTIGATOR ANSWERS

"WHERE"

HOW DO INVESTIGATOR ANSWERS

"WHY"

HOW DO THE INVESTIGATOR EXPLAINS THE SIGNIFICANCE OF HUMAN ECOLOGY?

ACADEMIC UNIT

Centre for Social Justice and Empowerment

(Planning and Evaluation and Quality outreach)

Outreach project "Training in farm, off-farm and non-farm sectors for human resource development with special reference to Tribal women".

Projects for : Research and training for human resource development in on-farm, off-farm and non farm sectors.

Institutionalisation of farm, off-farm and non-farm sectors for :

— Case Studies on Conservation of Ecosystems

— Studies on Women, Agriculture and Rural Extension with special reference of Biodiversity.

Project Summary

To conduct research towards improving the sustainable productivity and to support rural development sector, the project on women agriculture and rural extension will adhere to the NGOs through the collaborative research and training. The projects will contribute to the institutional strengthening of voluntary private sector through the development and delivery of packages for playing significant part both as catalyst and prime mover in rural development. The main thrust will be directed towards youth employment, income generation and generation of mass awareness about development issues. The activities which are socially beneficial and economically viable like on-farm and on-farm will be taken up for the benefit of the community by doorstep and on-station technology transfer through short term training programme to ensure continuous benefits to the rural poor as that the resource and environment may be preserved for future generations.

INSTITUTIONALISATION OF FARM, OFF-FARM AND NON FARM ACTIVITIES

FOR

SOCIAL JUSTICE AND EMPOWERMENT FOUNDATION MISSION

To help people, help themselves through the practical application of knowledge and resource to improve "their" Quality of life and that of future generation

Through

International Centre of Human Ecology

Academic unit

For Social Justice and Empowerment

For

(planning, evaluation and quality outreach)

DRIVER-PRODUCTION AND EMPLOYMENT

TRAINING	TARGET GROUP	TARGET AREA	OUTCOME
Pre Service	Students	Bangladesh	* Good supply of Extension Worker
	Managers	Bhutan	* Food Security — appropriate technology for agro based industries
Degrees	Extension Workers	Burma	
Ph.D.	Govt. Organisations	China	
M.Sc. (2 yrs.)	Ministries	India	* Sustainability in agriculture, dairy, horticulture, tiny & cottage industries and other agrowaste based income generating activities
		Japan	
		Korea	
		Maldives	
Diplomas		Nepal	
Post Graduate		Pakistan	
Diploma (1 yr.)		Philippines	
		Sri Lanka	
In-Service (3 months)		Thailand	
Short Term (2 weeks)			
Doorstep			
On station			

PROJECT : RESEARCH AND TRAINING FOR HUMAN RESOURCE DEVELOPMENT IN ON-FARM, OFF-FARM AND NON-FARM SECTORS

Awareness programmes and education cannot be seen in isolation from overall development programmes. These programmes should be directly linked to poverty alleviation programmes. It is very important to transfer the technology for appropriate activities through training programmes and to link up both at formal and informal level of education. Women play an important role in both education and development.

Through the years, a remarkable evolution has occurred as colleges of Home Science have responded to new issues and problems for making research a more visible component in the undergraduate programme; for increasing the sophistication and quality of graduate and faculty research and in encouraging a large number of faculty to engage in the advancement of both quality and quantity of research with the colleges of Home Science focus.

To provide social justice and empowerment especially to the disadvantaged sections for prompting the educational and economic interests of the weaker sections of the people including Scheduled Castes and Scheduled Tribes and protecting them from social injustice and all forms of exploitation. A wide variety of training are being offered in the field of extension from grass-root level upto Ph.D. degree. The programmes related to grass-root technologies need advancement and promotion. Delhi University had created a new centre for Human Ecology. This centre works for coalition of action in the areas like farming systems, biotechnology, food and nutrition and child care in particular and human ecology as a whole.

The future of the newly created centre, has an extremely important role to play and to marshal in relation to available resources and in controlling a broad range of environmental issues which deeply affect the quality of life of all.

We hope this venture will continue its landmark traditions of programme excellence and remain a catalysts for runner of "systemic thinking". Our vision and efforts will exploit the boundless opportunities if global technological challenges as we meet the need of individuals and families in traditional society with the inputs of established national and international education research.

TECHNICAL PROGRAMME (FOR SETTING PRIORITIES)

A. Food Security

- Sustainable management of fragile ecosystem such as highland and mountain areas, arid and semi arid regions, rainforests
- Linking commodity chains for production to marketing and consumption to ensure sustainable use of food resource
- Urban and peri urban agriculture.

B. Equity in Natural Resource Use

- Resource management policy in the country
- Decision-making processes that integrate environment, social and economic objects.

C. Biodiversity Conservation

- Market based incentives for sustainable use of the products of biodiversity
- Sustainable use of biodiversity to conserve plants
- Involvement of communities in development of agriculture and aquatic biodiversity.

D. Sustainable Employment

- The policy environment for job growth and options in trade policy and labour market policies
- Opportunities and challenges facing small and medium size enterprises strategies to promote innovations and improve environment management
- The repercussions of technological change in employment with special reference to communication technologies and biotechnologies.

E. Information, Communication and Evaluation

- The project evaluation to "what works" in development research activities. The evaluation process will be linked to decision-making policy formulation and programme management

- Knowledge access and transfer
- Networking with the use of information and communication technologies.

The training programmes :

Short term training programmes
Doorstep training programmes
Popularisation of biotechnology
Seminar, Symposia, conferences
Setting up units support for self-employment
Skill bank

The project is divided into three sections :

1. Administration
2. Research and Development
3. Training Programme.

GRANT-IN-AID SCHEME AND TRAINING FOR DEPRIVED OF INTERNATIONAL LEVEL

(Nepal, Bangladesh, Bhutan, Sri Lanka)

A. Research and Development (Formal and Informal Research)

Vision & Mission

This section will control the research on the following lines :

- Improve, Access to quality education and experts knowledge
- Achieve more active learning
- Generate new knowledge and scholarships across the mission
- Promote problem-solving to address society
- Advance diversity within community.

Control on :

Phase I

1. Development of training programmes and field attachment
2. Standardisation, implementation of training programmes

Phase II

3. Generation of skill banks for
   (a) market, (b) loan, (c) advance training centres
4. Evaluation of training programmes, modification and sustainability

Phase III

5. Development of course material — l year Postgraduate Diploma (a) compendium, (b) extension bulletins, handouts, pamphlets, poster for popularisation and replication, report writing.

Research and outreach during the past twenty years presents a dynamic picture of the vitality, diversity and significances of these two functions of human ecology. Outreach programme demonstrate flexibility in adapting approaches, methods, venues and emphases in order to serve people where they live with content and programming to meet their needs.

The centre for Human Ecology will continue to work towards the goal — "research-active or scholarly active and student focused" along with other units for the integration between and among research, outreach.

The faculty will provide outreach including talks to community, and professional groups, consultation with agencies, and organisations and membership on advisory committee to Govt. research bodies and other institutions.

Newspaper, Radio, T.V. will continue to be an out research project.

PURPOSE

To assist University Departments/Research Institute Professional bodies/Individual Scholars for conducting : (i) Research including action research; (ii) Workshop/Seminars; and (iii) Publication of Studies and documentation.

Science and Technology can be utilised to upgrade traditional skills, introduce new opportunities of work, reduce drudgeries faced during various household chores, improve the environmental and living conditions, reduce health problem. In fact, Science and Technology is capable enough of solving many problems of the people, if applied amidst their own environmental and cultural contact without creating any social and economic imbalances.

Environmental protection and development are the two facts of the same major issue. Environment and development are interdependent and cannot be separated. The questions related to environment are closely linked to human economic and social activities. The development of mankind cannot be sustained if problems of resources and environment are neglected. Likewise, protection of environment in isolation from economic development and technological progress has no basis. Therefore, we must trace back to the root cause of environmental degradation and find the best way to protect environment integrating it into economic and social development.

To assist those reputed social science research agencies which have a proven expertise and willing to undertake purposeful studies on the economic development, problems and requirements of Scheduled Castes and also on the impact of the implementation of various economic programmes undertaken by the governmental bodies through instruments of economic development.

To assist and develop linkages with governmental organisation for undertaking research including evaluation studies/projects, workshops and seminars helpful in orienting developmental programmes and disseminating knowledge and exchange of experience and publication of literature on tribal development.

OBJECTIVES

Institutionalisation on sustainable basis — The training programme on-farm, off-farm and non-farm sectors for human resource development.

Linking the curriculum from grass-roots to university level for commercialisation and management.

FOR THE WELFARE OF DEPRIVED (SCHEDULE CASTE, SCHEDULE TRIBES AND DISABLED) TO IMPROVE THE EDUCATIONAL, AND SOCIO-ECONOMIC CONDITIONS THROUGH TRAINING PROGRAMMES

B. Training Complex

A mid-career training for professionals involved in home-economics, nutrition or primary health care extension focusing in particular on the productive role of women in agriculture.

Farm Unit

No. of the trainings : 144

No. of the trainees : 1440 (10 per batch)

Duration : 3 months and 1-3 weeks short term courses

1 year Diploma Course in Women Agriculture, Rural Extension, Biodiversity and Environment

Tailor made programme for schools — teachers and students

Activities

(a) Kitchen Garden
(b) Fruit culture
(c) Mushroom cultivation
(d) Poultry keeping
(e) Nursery development
(f) Floriculture
(g) Cut flower industry — Wet and dry
(h) Prorogation techniques
(i) Microbes, insect, pest control

Total Activities — 9.

OFF-FARM UNIT

(Human Nutrition and Foods)

No. of the trainings : 112

No. of the trainees : 1120 (10 per batch)

Duration : 3 months and 1-3 weeks short term courses

1 year Post Graduate Diploma in Programmes for carriers for industrial research for govt. services.

Courses on patent law, environmental studies, professional health science and public health and food safety.

Activities

(a) Sprouts, fermented food items
(b) Indian sweets, snacks
(c) Bakery-cakes, pastries, biscuits

(d) Low cost-nutritious preparations
(e) Preservation of fruits and vegetables at home scale
(f) Meat, Fish and poultry — preparation and preservation
(g) Setting up of joints, restaurants, bar be que units
(h) Food safety — Public health (protecting the unsafe meat, poultry, dairy and sea food.

Total activities — 8.

Non Recurring

Equipment

Refrigerators, freezer, oven, microwave, gas connection, vessels, air conditioner, furniture, crockery.

NON-FARM UNIT
(Human Environment)

No. of the trainings : 160

No. of the trainees : 1600 (10 per batch)

Duration : 3 months and 1-3 weeks short term courses

1 year Post Graduate Diploma in Occupational Therapy.

(Tasks to restore, reinforce and enhance performance facilitate learning of those skills and functions essential for adaptation and productivity, correct pathology and to promote and maintain health for active role in improving health care and overcoming social problems.)

For the employment — In hospitals, Community Health Centres, Nursing Homes, Day Care Centres, Rehabilitation Centres, City Planning commissions, Half Way Houses, Home Health Agencies and Industries and Role of advocate for state legislation that affect the health care.

Activities

Domestic System

(a) Water system
(b) Electricity system
(c) Plumbing
(d) Drainage system

(e) Garbage system
(f) Recycling system — composting, vermi compost and recycling.

Cloth and Clothes Unit

(a) Dyeing, printing, screen printing, block, printing, tie and dye and batik
(b) Ironing, washing, finishing, drycleaning and setting up drycleaning unit
(c) Washing powder, soap cakes and detergent making, liquid soap, shampoos and stain removers
(d) Needle work — darning, embroidery.

Physical and Physiological Fitness Unit

(a) Beauty clinic and setting up of parlour
(b) Physiological fitness — Massage, Accupressure, Acupuncture, Yoga culture, Religion — Meditation.

Duplication, Word Processing and Information Unit

(a) Cyclostyling
(b) Photocopy
(c) Photography
(d) Word processing — typing, shorthand, printing and composing and book binding.

Information Unit

(a) Computer training
(b) Networking technology

Number Work

This programme will be incorporated with all the training programmes as per the duration of each programme

1. Economic Analysis — Production and pricing
2. Statistical Analysis
3. Quantitative techniques — decision-making process
4. Management accounting
5. Applied Economics — project planning and review Human Resource Development
6. Marketing — Marketing research, marketing management, Marketing communication

7. Business data processing :

   Computer concept, and programming, system analysis and cobol programming
8. Public Enterprise :

   Small business entrepreneurship, small business management.
9. Taxation :

   Principles and practices of taxation, corporate tax management
10. Corporate and Industrial laws.

Project Planning and Review

To impart necessary skills and evolve a suitable framework for the preparation, appraisal and monitoring of small business entrepreneurship.

With the focus on policy issue :

Project formulation — technical feasibility, estimation of costs, Demand analysis and commercial availability. Risk analysis collaboration arrangement.

Financial planning estimation and source of funds tax factors in project preparation, legal aspects of the project.

Project Appraisal

Business criteria — growth, liquidity and profitability. Social benefits/cost analysis in public private sectors, investment criteria and choice of techniques.

Project review : Evaluation of project, review of policies and programmes at the firm, industry and national level — a synoptic perspective case studies and field studies or other methods may be used to supplement lecture for this course.

12

NODAL CENTRE HUMAN ECOLOGY FOR EDUCATION

NODAL CENTRE HUMAN ECOLOGY FOR EDUCATION
(Planning, programming, monitoring and evaluation)

Financial Mechanism

For the projects for the development of infrastructure for *Relevance of education *Promoting of excellence and quality in education *Equity and access for management of higher education and UGC administration for "Financial Mechanism".

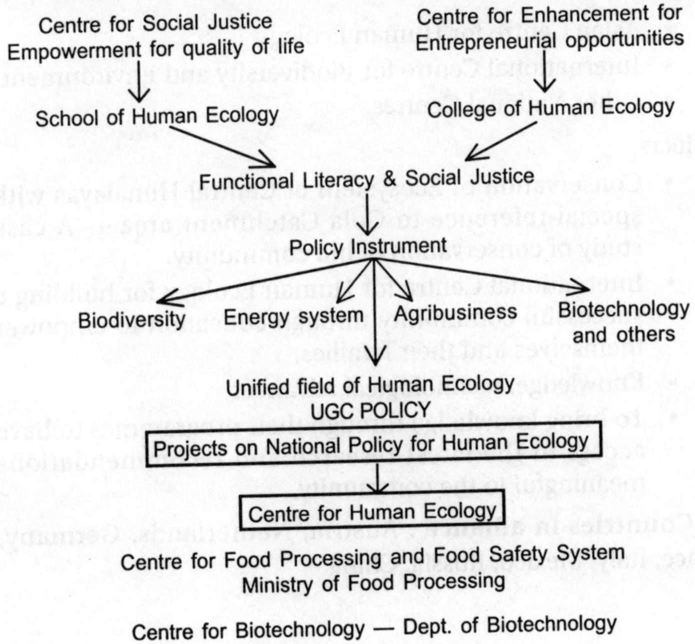

INTERNATIONAL CENTRE FOR BIODIVERSITY AND ENVIRONMENT

Providing education programmes to respond to human needs by improving nutrition and health, advancing environmental design and technology, promoting human development and securing economic and social well-being for individuals.

Drivers

Rural Development, Human Resource Development, Promotion of educational and economic interest for all sections of society, Multinational collaborations for building, scientific and academic community by linking with partners, Networking to increase and contribute to International research.

It will combine the interests, theories and studies of :

Geographers, Ecologists, Anthropologists, Home Economists/consumer Science Specialists, Psychologists, Agriculture Scientists, Economists and Biologist.

Centres

- Asian Centre for Human Ecology
- International Centre for Biodiversity and Environment
- Other National Centres.

Projects

- Conservation of Ecosystem of Central Himalayas with special reference to Gola Catchment area — A case study of conservation of hill community.
- International Centre for Human Ecology for building a successful community through education to empower themselves and their families.
- Knowledge, technological advances.
- To bring knowledge through their programmes to have access to the latest research and recommendations meaningful to the community.

Countries in alliance : Austria, Netherlands, Germany, France, Italy, Mexico, Russia, China.

(FOR LINKING HOME SCIENCE CURRICULUM WITH FARM WOMEN AND COMMUNITY THROUGH STUDIES ON WOMAN AGRICULTURE RURAL EXTENSION.)

School of Human Ecology	College of Human Ecology
• To cover home science curriculum Under NCERT system	* Deptt. of Family & Child Ecology
	* Deptt. of Food Science & Human Nutrition
	* Deptt. of Human Environment
	* Deptt. of Psychology
	* Deptt. of Social Sciences

Fund support for the project : "Conservation of ecosystem of central Himalayas with special reference to Gola Catchment area" (A case study of conservation of ecosystem of hill community).

Drivers

- To provide various groups of Human Ecology for women (as well as the who are to be trained in imparting and in developing the field of Human Ecology) and consequent, challenges/opportunities for teaching education in the field of human ecology.
- To review currently employed teaching methods and assess how far these are suitable, and what appropriate methods and training programmes need to be devised.
- To devise programmes and techniques for sensitising not only the institutions (the teaching faculty and students body) to imbibe human rights culture but also for developing sensitivity and respect for human rights at the community level.
- To identify activity and programmes in addition to running courses, that could be undertaken by the institutions for objectives of human ecology education.

Establishment : UP HILLS — Centre for Human Ecology School of Human Ecology for Hill Community.

ESTABLISHMENT OF NODAL CENTRE FOR HUMAN ECOLOGY

(*Society and science provide knowledge to create human ecology. Its interactions between human and their environment is in making the technical concepts more realistic and meaningful.*)

To establish Nodal Centre for Human Ecology, with the view of present education system, the criteria for distribution of funds to the institutes of higher learning and research by a parameter of productivity of the institutes rather than their products and image etc., and to conduct adult continuing education and out-reach activities, the department/centres should become an integral part of all departments and constituents of the University through the optimum involvement of teachers and students. The universities need to give a statutory status to the departments with the core faculty eligible for representatives on all the university bodies such as senate, academic council, etc. To establish the Nodal centre for Human Ecology at JNU for Research, Education, Extension and field outreach for the optimum involvement of teachers and students where Centres will be the integral part of academic work. The basic aim of the forwarded projects is to adopt an integrated approach with no "hidden hunger" for ecological access to our population below poverty line particularly children women suffering from mal-nutrition and go to bed hungry every night.

All these reflect the pressure to produce more diversified food at much accelerated pace. Widening regional disparity in agricultural development is also a major concern and need to be addressed intervention and extension strategies are required to play a role in poverty elevation to provide nutritional security for the poor. Appropriate research intervention and extension strategies are required to be developed to make such allied sector more productive and vibrant partnership.

To accomplish above task, process of research reprioritisation and consolidation and redeployment of resources along with modernisation, JNU has infrastructure and human resources for better efficiency and output. The strategy of *panch sutras* — people, productivity, permanency, policy and partnership, however, will be central in attaining the goals, food, and environmental security.

University Profile

Name of the University : Jawahar Lal Nehru University

State : Delhi

Date of Establishment : 1970

Name of Vice Chancellor : Prof. Asis Datta.

Schools in the University

School of Computer and System Sciences
School of Environmental Sciences
School of International Studies
School of Language, Literature and Cultural Studies
School of Life Sciences
School of Social Sciences.

Several centres in these schools have been declared by the UGC to be "Centres of Excellence". These Centres for Historical studies, Centre for the Study of Social Systems, Centre of Political Studies, Centre of Economic Studies and Planning, Centre for Study of Regional Development.

INNOVATION OF THE DISCOVERY OF KNOWLEDGE FOR COMMUNITY BASED APPROACHES

To develop unique scientific culture by linking several disciplines and professionals for ecology of individuals, families in their diverse socio-cultural, physical, biological environment where diversity of ecosystem is way of life.

To work with the Centres of JNU for organising pre-service and in-service courses, refreshers training with the centres of excellence like Zakir Husain Centre of Educational Studies, School of Social Sciences, School of Life Sciences, School of Physical Sciences, School of Environmental Sciences and Centre for Biotechnology.

For action to be taken in order to overcome institutional bottlenecks to participatory programmes.

- For the sustained process of endogenous innovation generating activities as indicator of success for continuity of learning in academic structure for the involvement of curriculum work and re-animating the collaborative process of generating knowledge in the classroom and taking independent research by the students.

To develop In-House Research and Development Activities for untrodden path to the journey of Human Ecology requires the support in order to understand the process of development, various institutional technological and environment factors of different regions. For the development of human resources in the area of biotechnology for the exploitation in the field of industry, medicine and agriculture. For teaching physico-chemical aspects of biology, cell biology, bio-chemistry, bio-physics, microbiology, molecular biology, bio-statistics and computer applications.

- Instrumentation CIF (Central Instrumentation facilities)
- USIC (University Science Instrumentation Centre)
- For Rich Soft Ware Library
- For the Study of Law and Governance.

Basic concepts — Paradigm in shifts in human services that are promoting change for better outcomes for children, families, neighbourhoods and communities processes and tools — ways of operating that can facilitate the new way of doing business.

Best practice — models — Tried and tested human service delivery basic research and evaluation reports and field experience the best information currently available than decision makers can use in human service programme development.

Education programmes — to over a wide range of diverse topic, programming efforts to address community based issues in many areas including agriculture, natural resources, community and economic development families and children.

Children youth and family : programme to use the educational resources and methods of following areas to deliver educational programmes within community to enhance the quality of life across the life cycle.

Family strength/home economic

Food nutrition and health

Youth development

Training programmes in horticulture education, location

specific training programmes by doorstep technology transfer and on station technology transfer.

Development Area of expertise team in Family strength/ Home economics — Area of expertise team to help families, succeeded by integrated university and community resources in educational process that applies, knowledge, practice application of research in critical issues.

Field crops — To help crop producers, improve themselves, their business and their environment through an educational process that applies knowledge demonstration and research to critical issues of biology, cell biology, bio-chemistry, bio-physics, microbiology, molecular biology, bio-statistics and computer applications.

- Instrumentation CIF (Central Instrumentation Facilities)
- USIC (University Science Instrumentation Centre)
- For Rich Soft Ware Library
- For the Study of Law and Governance.

Basic concepts — Paradigm in shifts in human services that are promoting change for better outcomes for children, families, neighbourhoods and communities processes and tools — ways of operating that can facilitate the new way of doing business.

Best practice — models — Tried and tested human service delivery basic research and evaluation reports and field experience the best information currently available than decision makers can use in human service programme development.

EDUCATION PROGRAMME

Human Ecology combines the interests theories and studies of geographers ecologists and anthropologists with the focus on the modification of the earth's ecosystems. Ecology and environment links botanists, plant ecologists, conservation biologists, resource managers, and students at the University or high school levels. The colleges of Home Science are to respond to human needs by improving nutrition and health, advancing environmental design and technology, promoting human development and securing economic and social well-being for individual.

Human ecology considers human affairs as :

- Human factors in their evolutionary and ecology context as part of a way of looking at the world and its animals, plant, natural resources, its geography and its geopolitics for human behaviour studies.
- Engineering and technology, industrial design of industrial hygiene, environment and development, microbial ecology, to improve the practices and educational standard.
- Education for environmental sustainability and social justice, working with communities, business and students to develop consciousness, professional skills, creativity and leadership.

Extension programmes cover a wide range of diverse topics. Programming efforts address community-based issues in many area including agriculture, natural resources, community and economic development, families and children. These are :

Under Graduate and Graduate Programmes

Children, Youth Families

- Children, Youth and Family programmes use the educational resources and methods of the following areas to deliver educational programmes within communities to enhance the quality of life across the life cycle :
- Family strengths/Home Economics
- Food, Nutrition and Health
- Youth Development
- Integrated Pest Management
- Integrated Pest Management is a suitable approach for managing insects, weeds and plant diseases that combines the use of biological control, cultural, physical and chemical tactics in a way that minimises economic, health and environmental risks.

Academic Programme for Resources

- Family and Child Ecology
- Food Science and Human Nutrition
- Human Environment and Design
- Institute for children, youth and families

- Distance learning programme
- Community and Rural Development programme
- Case studies to research extension and teaching in support of community and rural development programme
- Resource links to community development-related programmes and projects
- Community development conference and training events for various audiences throughout the year
- Community development projects as experimental education for rural people with limited income
- Agriculture and farmland protection action kit
- Approaches to rural research connecting research and practice with rural sociology and other rural studies to share subject matter.

Area of Expertise

The faculty for human ecology will combine expertise in interior and product design, housing architecture, planning, environmental psychology, human factors and ergonomics creates a unique learning environment at the forefront of design, facilities planning and management, and human environment relations.

Area of Expertise teams — to involve Extension specialists, agents, researchers and/or stakeholders organised around a particular commodity, interest area and/or issue. They will be charged for listening to stakeholders, identifying priorities, planning and providing educational programmes, and evaluating programme outcomes and impacts in the following areas :

Family Strengths/Home Economics

The family strengths/Home Economics AOE Team helps families succeed by integrating university and community resources in an educational process that applied knowledge, practical application and research to critical issues.

Conservation Biology

Biodiversity, ecology and environment with global changes

and ecological research network for ecosystem, inventory, monitoring and long-term ecology research.

Field Crops

The field crops AOE team — to help crop producers improve themselves, their business, and their environment through an educational process and applies knowledge, demonstration and research to critical issues.

Firm

The mission of the Agribusiness Management Educators to assist agriculturally-related industries, business, families and individuals to improve their lives and achieve their goals through an educational process that applies management and economic knowledge to critical issues, needs and opportunities.

Food, Nutrition and Health

The food nutrition and health AOE teams for sound food, nutrition and health principles to help families improve health through an educational process that applies knowledge, practical application and research to critical; issues (Food Safety, Forestry).

Tourism

The tourism AOE team to help the tourism industry to improve themselves, their business and their environment through an educational process that applies knowledge, demonstration and research to critical issues.

Youth Development

The youth development — to positively develop young people and the volunteers and professionals who work with them through an educational process that applied knowledge, practical application and research to critical issues.

- The Crop Advisory Team (CAT) newsletters and handouts to feature timely pest management information based on current conditions. Faculty and extension field staff to write the reports and articles that address both immediate and long-term production and pest management concerns.

Mission

The office for research, outreach and international programme with the purpose of facilitating and supporting home science/home economics college faculty scholarships in research, outreach and international programmes. It will be a resource point of access, for faculty in focusing enhancing the generation and application of knowledge across the domains of research out-reach and international programmes.

- The office will be committed to the mission of the colleges of home science of empowering and strengthening families and individuals in address the challenges of and opportunities in their environment.
- Through the support of research, outreach programmes, the office shall promote scholarship and initiatives that enhance the positive development in communities and acknowledge the diversity of families and community served by the colleges.
- The office will encourage scholarly efforts across the broad mission of JNU and its centres.

Goal

- It will support the collaboration of home science faculty in the persuit of research priorities which cuts across the three departments.
- It will promote, facilitate community outreach which reflects and embodies the influence of college research and scholarship programmes for people, business and industry.
- To initiate public policy efforts which address the needs of families and communities.
- It will facilitate the development of international programme which involve faculty in collaborative research and outreach/research with faculty scholars, in the countries around the world.

Selected Strategies

- Encourage the development of faculty research proposals
- Provide seed grant funding for faculty research that will be used as foundation, and leverage to obtain external funding

- Support proposal writing and grants manship seminars in conjunction
- Link faculty with collaborators to support in grant acquisition process
- Work closely with university administrations and faculty, individuals, on specific research interest of the faculty as well as outreach and international programmes
- Organise and develop faculty research groups in relation to identified research themes.

13

Centre for Human Ecology for Urban, Slum Communities of Delhi and Neighbouring States (Delhi Based)

WORKING WITH COMMUNITY

In our experience, what bring success are not just good weather and good grain prices, but strength of character in response to crises and opportunity to develop such traits. In today's vocabulary, we can describe some of these as personal responsibility, the courage to act; a respect for diverse voices; and civics participation. These attributes are the hallmark of healthy people, strong families and productive communities. As a foundation, we must always evolve in response to social change. This is the way of any healthy organisation. What will not change, however, is our belief in the power of committed people, with valid ideas, and sound character, to forever improve the world they live in.

As we approach the new century, people around the world see a similar era of discovery. But many of the frontiers we face today are personal and social. They tie within us, within our families, within our communities. They include new ways of thinking, working, healing, and living that hold great promise for humanity. Our machines and computers, powerful as they may be, possess not a wit of spiritual wisdom or enlightened judgment.

To yield stiff fiber to be used for cordage and rope — impoverished farmers lacked the land development resources to improve their quality of life. To bring ahead today for the reconstruction of small farmers with solar energy systems, to provide food and fodder plants and "their training leaders, to guide farm families towards sound nutrition and scientific agriculture and animal husbandry, with the host of community partners. To establish the agriculture family school to offer a flexible :

- Basic education grounded in rural reality with the new resources today, training for tomorrow and hope for the future of its youth for beginning to bloom.
- Institution — towards a world of respect for self and others for growing closer today's younger generation — distanced by curriculum of stereo type in popular culture.
- In the face accurate, update information via computers, and traditional learning materials based on in truth and wrapped in goodwill to foster self-model for educators in other lands.
- To train them along, salutary lines and to hold them under control, certain good discipline. Until the taste sprouts, the discipline have to be strictly followed.
- For constant practice and creation of zest.
- For the taste for such wholesome literature for accumulation of literature out of merit.
- For listening to exposition by wide — is much better than reading them, in company, rather than alone (of course, it is excellent to listen with a number of earnest aspirants).

(for the profit that one can derive while in the company of great : for shining with genuine light for the fulfilment of his destiny).

INTRODUCTION

Shared vision on food and nutrition and environmental security, the major problems of food production and their efforts are diverted towards poverty alleviation, achieving economic growth and natural resource management. Research and education have helped various nations in meeting their food needs. However, commensurate to UGC, Resource crunch and the problems of organisation and management effecting adversely their effectiveness and efficiency. Required attention is therefore, needed to build management skills to improve and strengthen output and efficiency, being needed organisation and management reforms to have better efficiency and visibility, the need for human resource development, information system development policy planning, monitoring, research prioritisation, linkage with international scientific institutions, complementary role of public and private sectors, resource generation mobilisation and efficient utilisation of available resources, more involvement of stakeholders in research governance priority setting, research execution and evaluation, innovation in technology transfer involving farmers, development departments, NGO's private sector in a variety of formal and informal ways such as Research — extension — farmer linkages. To meet the emerging requirements of agricultural research, education and development for strengthening centre for Human Ecology at University of Delhi with the vision as fully autonomous, efficient, effective caring and financially sustainable organisation for achieving the goals of household security food and nutrition security economic welfare and sustainable use of natural resource. Following project proposal have been submitted to strengthen in house exercise to bring in needed operational and management reform. The infrastructure at the Institute of Home Economics need to be modernised and competent human resource is required to manage both research and management. Increased research investment and accountability is necessary to accrue the benefit from research.

It is my fond hope that this centre shall help individuals and agencies involved in development work. And activism in realising the dream for better eco-friendly humane and sustainable society by travelling on the road which is less travelled.

REGIONAL PRIORITIES IN RESEARCH CAPACITY BUILDING AND LINKAGES

New Thrust and future planning for :

- Expansion of research and outreach with people resource and facilities
- Postgraduate degree/diploma programmes for pre-service, in-service graduates

- Other training programmes — short-term and doorstep and tailor made programmes as per Need.

Objectives

To formulate the Home Science curriculum for conducting studies on women agriculture rural extension through education and training programmes with special reference to pre-service graduates :

— To fight poverty with passion and professionalism for lasting results.

— To help people, help themselves and their environment by providing resources, sharing knowledge, building capacity and foreign partnerships in public and private sectors.

— To be an excellent institution able to attract excite and nurture diverse and committed staff with exceptional skills who know how to listen and learn.

Joint project — can be linked with National level similar kind of projects related to home science curriculum for human resource development with other developing countries in health, food systems and rural development.

- Youth education/higher education
- Integrated programmes
- Cross cutting themes/strategic themes — leadership.

Information system and technology, social economic community development.

Objectives of the Programme

- To meet the challenges on the critical issues including financial constraints, monitoring of resources, Management capacity and related overall as facilitator in regional coordination and planning.
- Setting strategic directions for priority issues : operating environment, priority setting and strategic planning and financing.
- Network initiative.
- To monitor research and development to meet emerging needs; to review priorities on a regular basis to take into account the new information/development

and changes in operating environment/organisational management.

- To involve private sectors in contact research to achieve time frame research results.
- Capacity building for strategic planning to increase their efficiency and effectiveness.
- Overall to mobilise support for agricultural research from high level makers, donors etc., at regional level as a facilitator and promote coordination/collaboration in R&D activities.
- The project "Studies on women agriculture rural extension and training programme with special reference to pre-service graduates" for high priority activity at national and regional level. The existing national capacities/capabilities need to be assessed and special training needs which may be common to sub-regions/regions or specific to the centre for human ecology.
- To identify problem and needs Development of technology in particular mode *i.e.*, Research Extension workers — Farmers — links between research and extension is very important for the transfer of desired technology.
- Use of several tools/measures such as demonstration technology, pilot production programme, lab to land programmes with multimedia audio visual aids, NGO meetings; field days etc.
- The project proposal for research management mechanism; for resolving poverty food and nutrition security issues. Conservation and management of national resources for sustainable agriculture and rural development.

Drivers

- To provide various groups of Human Ecology for women (as well as the who are to be trained in imparting and in developing the field of Human Ecology) and consequent, challenges/opportunities for teaching education in the field of human ecology.

- To review currently employed teaching methods and assess how far these are suitable, and what appropriate methods and training programmes need to be devised.
- To devise programmes and techniques for sensitising not only the institutions (the teaching faculty and students body) to imbibe human rights culture but also for developing sensitivity and respect for human rights at the community level.
- To identify activity and programmes in addition to running courses, that could be undertaken by the institutions for objectives of human ecology education.